GENES AND THE BIOIMAGINARY

Genes and the Bioimaginary

Science, Spectacle, Culture

DEBORAH LYNN STEINBERG
University of Warwick, UK

ASHGATE

Published by
Ashgate Publishing Limited
Wey Court East
Union Road
Farnham
Surrey, GU9 7PT
England

Ashgate Publishing Company
110 Cherry Street
Suite 3-1
Burlington, VT 05401-3818
USA

www.ashgate.com

British Library Cataloguing in Publication Data
A catalogue record for this book is available from the British Library

The Library of Congress has cataloged the printed edition as follows:
Steinberg, Deborah Lynn.
   Genes and the bioimaginary : science, spectacle, culture / by Deborah Lynn Steinberg.
      pages cm
   Includes bibliographical references and index.
   ISBN 978-1-4094-6255-2 (hardback) – ISBN 978-1-4094-6256-9 (ebook) –
   ISBN 978-1-4094-6257-6 (epub) 1. Human genetics–Social aspects. I. Title.

QH438.7.S725 2015
304.5–dc23

2014042725

ISBN 9781409462552 (hbk)
ISBN 9781409462569 (ebk – PDF)
ISBN 9781409462576 (ebk – ePUB)

Printed in the United Kingdom by Henry Ling Limited, at the Dorset Press, Dorchester, DT1 1HD

Contents

Acknowledgements

I owe a considerable debt of gratitude to many for their support of this book. Thank you to: Margaret Archer; Avtar Brah; Mariette Claire; Ron Davis; Gillian Einstein; Tony Elger; Debbie Epstein; Elizabeth Ettorre; Sander Gilman; Dorothy Sandler-Glick; Chris Griffin; Samantha Halliday; Gillian Lewondo Hundt; Richard Johnson; Mary Jane Kehily; Beau Lamour; Nina Lykke; Maureen McNeil; Jane Monger; Lyndsey Moon; Stuart Murray; Ian Proctor; Eleni Prodromou; Rayna Rapp; Peter Redman; Barbara Katz Rothman; Rivanne Sandler; Sari Simkins; Margrit Shildrick; Gershon Silins; Meg Stacey (now, sadly passed away); Maxine Steinberg; Irwin Steinberg; Lucy Suchman; Jonny Wallis; and Simon Williams.

I would like to thank the Centre for Women's and Gender Studies, University of Toronto where I was based as a Visiting Professor in 2006; participants and organisers of the Technoscience Salon, University of Toronto; Michelle Murphy, Brian Beaton and participants in the Biopolitics Workshop, University of Toronto (where I was able to develop my early ideas for Chapter 7 on phantasmatics and the search for the Jew's gene); Gillian Lewondo Hundt, guest hosts and participants in our ESRC seminar series 'Cultures of the Gene: Discourses, Dialogues, Debates', Dorothy Sandler-Glick and participants of her Toronto Salon; colleagues at the Open University, Department of Sociology and the ESRC seminar series organised by Simon Williams (where I presented early versions of Chapter 6); The post-CCCS Narrative Group (in which I was able to extend and refine the methodological approaches that inform much of this book); Elizabeth Ettorre and participants in the ESRC seminar series on 'The New Politics of Reproduction' and Nina Lykke and participants in the 'Feminist Research Conference: Gender Genes and European Biopolitics.' (where I presented work in progress on moral signification and the 'gay gene' that became Chapter 4); colleagues in the Department of Sociology at Warwick University who attended the (trial by fire) Department Research Seminar – especially Margaret Archer, Tony Elger and Ian Proctor (where I presented early ideas that became Chapter 3 on materiality, metaphor and the gene) and also colleagues who attended the Sociology Department Annual Debate between myself and Sander Gilman (and where I presented the first version of what became Chapter 8 on what is 'real' about genes). Thanks also to everyone who participated in the vote on the title. I would also like to thank my editor, Neil Jordan at Ashgate.

I owe particular gratitude to the 'Birmingham School' – the Centre for Contemporary Cultural Studies, University of Birmingham UK, ignominiously closed down by the University in 2002 – which I had the great good fortune to join in the formative years of my intellectual journey. I hope that this book helps keep that tradition moving forward and does justice to the friends and colleagues from the Centre that have played such a central role in my own love of this work.

Introduction

We are living in the age of the gene. There is no doubt that the gene, once the preserve, as object and artefact, of primarily scientific preoccupation, has now powerfully and pervasively infiltrated the full fabric of contemporary life. The language of genetics pervades our culture as the era of genetic engineering capabilities progresses through its fifth decade, and has heralded a revolution, not only in the arenas of biotechnology and medicine, but also across politics, popular culture, political economy and everyday vernacular.

Since the 1970s, developments in recombinant genetics have appeared at pace, heralding radical transformations in medical, industrial and agricultural practices; in regulatory relations of law, commerce and global development; and in commonsense understandings of risk, disease, kinship, citizenship and identity. Genetic science has posed unprecedented bioethical and regulatory challenges surrounding the meanings of intellectual property, social utility and risk. It has raised new questions and significant concerns about the powers of biomedicine and commercial interests *vis-à-vis* non-professionals, including patients and the general public. It has provoked anxieties and invited aspirations, has altered terms of reference and debate on race, gender, class and sexuality, and has presaged a radically reconfigured climate for reproductive politics and practice.

Perhaps most importantly, the possibilities projected onto the understanding of genes are imbricated not only in novel scientific and medical practices, but also in processes of signification and grand narrative. As this book will explore, the emergence of the gene in and as contemporary commonsense accrues from a complex authorship in which the authoritative accounts of scientific research intersect with the range of representational media – literary, press, digital, televisual and cinematic, lay and professional. The gene is thus, at one and the same time, a site of practice and of projective identification, configured as a tool and also as a symbol.

This book represents the culmination of over a decade and a half of observation and critical assessment of the dramatic rise and cultural apotheosis of the gene. The book traces not only the epochal 'genetification' of the culture but is also an intellectual journey that has aimed to make sense – through a focused examination of the rise and sedimentation of the gene – of the complex and intersectional relationship between science and culture. At the heart of this book are two interlinked questions. The first concerns the paradigmatic transformations heralded in the wake of the 'genetics revolution'. And second, what are the means by which we can gain critical purchase not only on the conditions and consequences of a particular science, but on its projective seductions, the terms of its persuasion, and the dilemmas and anxieties provoked in its wake?

More specifically, this book explores the pervasive and progressive cultural purchase of genetic discourse and the articulations of the gene between biological and cultural imaginaries. It forges a critical meditation not only on the epistemic, but also on the affective and phantasmatic dimensions of genes and genetics. It explores the ways in which the gene has emerged not only at the interstices of science *and* culture, but also as a product of science *as* culture. In other words, this book explores genes and genetic discourse as sites of a culture industry, and explores science as a site of cultural authorship. The book thus interrogates and challenges the notion that culture – and cultural questions such as feeling, spectacle, discourse, persuasion – is outside and epiphenomenal to the scientific enterprise. It forges an alternative understanding in which the privileging of scientific discourse is, itself, a cultural phenomenon.

The age of genes has witnessed profound transformations in the political, economic and regulatory dimensions of science, in the interface of technology, digitalisation and industrial practice, and, as I suggested above, in understandings of bodies and identities, health and illness, kinship and ethics. Taking a case-study approach and drawing on an interdisciplinary array of critical analytic tools – drawn from semiotics, cultural psychoanalysis, narrative studies, and feminist, media and cultural theory – this book explores the rise of the gene in three respects: i) as a site of knowledge production crossing boundaries between the clinical-scientific and the popular; ii) as a gateway technology and locus of transforming bioethical values and modes of bodily governance; and iii) as site of spectacle, projective fantasy and attachment.

Science, Spectacle and Desire

Central to this book are three methodological questions: First, by what means do we gain purchase on the power relations and seductions of biomedical discourse? Second, what are the processes we aim to theorise in order to explain the forms biopower may take and the knowledge economies through which the products and processes of biomedical practice are forged, regulated, reproduced (or resisted)? Third, what critical tools enable us to grasp both the anxieties and attractions that may attach to some sciences and not others?

Episteme

This book begins from the proposition that the spectacle of science is not simply an epiphenomenal artefact, tacked on to 'real science', but is, rather, part of the epistemic core of scientific cultures and scientific work. Thus, exploration of moments and contexts in which spectacle and science explicitly come together tells us something, indeed something crucial, about both. My concern here is with the dialogic relations between science and signification as they consolidate around the gene. The book aims to theorise the intersections of what Foucault (1966) termed the *episteme* (the conditions of possibility of knowledge), affective attachment (the conditions of knowledge's *plausibility*) and phantasmatic projection (the meta-narrational, symbolic investments that drive or accrue to scientific enterprises). I am interested in genetics as discourse and representation as well as practice.

To this end, the individual chapters of this book examine points of nexus between science and popular culture. Each chapter is organised around both substantive and methodological questions, offering differing critical vantage points from which to interrogate the interpretive-imperative field of genes. In turn, the juxtaposition of these cases suggests a wider gestalt, the broader bioimaginary, the underlying biopolitical and bioethical sensibilities of genes. The first half of the book is concerned with the genetic *episteme* in the original Foucauldian sense: that is as noted above, with the conditions of possibility (and consequences) of knowledge. Thus, Chapter 1 examines questions of biopower and gendered medico-morality in the context of reproductive (diagnostic) genetics. Chapters 2 and 3 draw on Sontag's insights concerning language and power in their examination of the dominant metaphors of genes and in particular in their (post)racial narrations and symbolic economies. Chapter 4 turns to modes of popular signification as they play out over the multi-layered moral terrains of gays, genes and governance. In Chapter 5, the book turns to its second consideration: the conditions of *plausibility* of knowledge. Beginning with the example of crime and justice, subsequent chapters progressively unpack the affective, seductive and phantasmatic dimensions of genes, examining not only the politics, but the persuasion of genes, their aesthetics, their purchase on the terrains of neoliberal ethics (Chapter 6), cultural identity (Chapter 7), what is 'real' (Chapter 8).

Each chapter pursues a distinctive methodology. In so doing, each provides a focused, parallel meditation on the purchase of particular critical concepts on the intersectional terrains of science and culture. The critical tools deployed in this book draw from social semiotics, film studies, cultural psychoanalysis and traditions of narrative and discourse analysis. The case studies range across biomedical texts, to popular spectacle and cross genres of scientific peer review, science reportage, documentary film, television drama and literary fiction. Taken together, these examples provide a powerful evocation of the wider cultural field of genes, providing a multivalent interrogation of contemporary late-modern meta-narratives as they have coalesced around the gene. These include biopower and subjectivity, capital and modes of bodily governance, 'risk society', the 'turn' to extremity, the primacy of spectacle and the postmodern dispersal of identity. Thus the socio-cultural, biopolitical and bioethical dimensions of genes articulate with the power of spectacle and projection and the cultural negotiations that constitute genetic capital. In this context, the book offers a significantly reworked paradigm for understanding the epistemic foundations of science. Here knowledge is regarded not as a purely (or even primarily) cognitive formation, artefact, or praxis, but also as a locus and effect of feeling, of attachment or repudiation, of aspiration, of revulsion. More specifically, this book argues that knowledge, among the many things that it may be said to constitute, is also a *persuasion*. It is a terrain of affect, of standpoint. To extrapolate from Eve Sedgwick (1993),[1] knowledge has *tendencies*.

In its 'turn to affect', the book offers a reworked and expansive interpretation of *episteme*, arguing that the conditions of possibility of knowledge emerge not only from the articulations of capillary and convergent power, but also from the conditions of its *plausibility*. Hence the foundational importance of genes as sites, not only of practice and authoritative knowledge, but also of discourse and spectacle, of feeling and desire. And hence the consideration of the authorial, as well as authoritative relations of science – of its role, in other words, *in* and *as* a culture industry.

Social Semiotics and the Gaze

In addition to its reworked notion of *episteme*, this book takes up a number of critical media-cultural theoretical concepts to unpack the gene as a site of *signification*. These include the notions *commonsense* (the common or presumed assumptions or 'truisms' of a culture), and *discourse*, referring not only to textual modes of signification, but, following Foucauldian tradition, to the broader relations knowledge-power that articulate in scientific practice, in its bodies of knowledge and in the repertoires of meaning that emerge from and accrue to it. The book is interested in both textual repertoires (language, metaphor, narrative) and visual-generic conventions and aesthetics. It draws liberally on Mulvey's notion 'the gaze', to unpack the subject-object relations of the gene as spectacle. Underpinning these approaches is a particular understanding of signification as embedded in and constitutive of (as opposed to simply 'representing') the social

---

1 Sedgwick's evocative use of the term 'tendencies' arose in her meditations on sexuality politics and identity and shaped the underpinning (postmodern) epistemology of queer studies. See also Arendt's (1963) discussion in *Eichmann in Jerusalem*, of the imperative freighting of knowledge and power and where she distinguishes between what is overdetermined (a tendency or persuasion – in this case the totalitarian regime-episteme of National Socialism) but not inevitable or totalising.

field. Thus, as I shall explore, the symbolic economies of the gene are imbricated in the material field (the social conditions and social consequences) of genetics.

Media Ecology of Genes

Immersion is not usually a concept associated with media studies, other than perhaps a study with an ethnographic element involving some form of embedding in a specific field.[2] And until the latter part of the production of this book, I did not consider myself as 'embedded' either in the ethnographic sense or as per common parlance in journalistic circles. And yet, embeddedness characterises both the field and the methodological approaches of this book. Each chapter takes up an immersive approach to its specific subject-matter, involving focused scrutiny and exposure to categories of text across genres from textbooks and scientific articles, popular texts, journalism and a widening range of visual and radio based media.

Serendipitous developments in digital media,[3] moreover, dramatically filled out these localised immersive case studies, highlighting (and leveraging) the symbolic capital and purchase genes within a vast and infiltrating *media ecology*. This book became progressively shaped by the development of opportunities within online news outlets to tag and archive categories of reportage and particularly sites that aggregate across digitalised sources (including news, magazines, scientific articles, blogs and other web-based publications and indeed other aggregating sites). Over the past two years in particular, I have experienced a degree of immersion across genres and media that I could not have imagined when I began the project. Thus, I was able to witness the occupation of genes not only in what Richard Johnson (1986–7)[4] termed a *circuit of culture* (the multi nodal processes – spanning production, consumption, representation and regulation – that produce genes as objects of knowledge/meaning), but the ways in which the meaning circuit of genes, is also, itself part of a globalising-insinuating and networked technology of knowledge that both troubles and produces distinctions between truth and desire, between knowledge and persuasion.

Fantasy/Phantasy

Finally, this book takes up an array of cultural psychoanalytic concepts to interrogate the affective field of genes, that is, the terms of *plausibility* of genetic

---

2 Ann Gray's (1992) ethnographic study of gender and video use in a domestic setting takes an embedded vantage point on leisure and media consumption.

3 The iPad's virtually seamless interface between email and Safari allowed me to integrate into my reading practices system of archiving that could easily amass and qualitatively catalogue an unprecedented volume and range of material.

4 See also Du Gay et al. (1997) and Miller et al. [The Glasgow Media Group] (1998).

knowledge. In particular, it aims to explore the inter-resonances of science, spectacle and modes of desire, and to re-theorise the gene as an elaboration not only of knowledge practices (scientific and popular) but, significantly, of 'regimes of feeling' and *phantasmatic* projection. Central to these regimes are the interlinked discursive-psychoanalytic questions of fantasy and phantasy. All of the case studies explored in this book take up modes of aspirational (or repudiative) fantasy – that is, imaginative projections, articulated in narrative form, of desired goods (or repudiated ills) that attach to genes and genetic science. Such projections are, in turn, leveraged through modes of feeling that speak to larger, more inchoate yearnings and anxieties – phantasy. This latter question – the phantasmatic underpinnings of genes – cross over genre and context, and as this book will explore, cement the inter-resonances of science and popular culture, of knowledge and persuasion, and of genetification and genetic governmentality. As we shall see, the gene is a point of considerable phantasmatic elaboration, crossing terrains of agency (phantasies of action), communality (of belonging) and (im)mortality (melancholia).

In this context, perhaps the most important mode of phantasy is what I would term *normotic phantasy*. Here I draw on, but also significantly depart from, Bollas' (1987) description of 'normotic illness' – that is, the idealisation and enactment of self as 'pathologically normal'. Normotic illness, Bollas suggested, is in part characterised by a totalising (compulsive) investment in the normative. Extrapolating from this, and taken as a cultural description, *normotic values* and *normotic phantasy* might describe the invested repudiation of what is subjective (feeling, the body) in favour of an idealised ultimate rationality. As such, it can be suggested that normotic phantasy constitutes the cultural "unconscious" or *persuasion* of modernity and modern science,[5] with their principled claims of rational utilitarianism, objectivity and natural law. The place of normotic phantasy in the context of genes, is a primary theme of this book, and spans the discussion of diagnostic governmentality (what Lemke (2004) has termed a *government of genes*) to the validation of the Lemba as Jews to the social activism of *The Innocence Project*.

Critical Studies of Genes and Genetics

Despite the pervasive infiltration of genetic discourse and genetic practice across the gamut of social and cultural life, critical explorations of the gene and

---

5 Or as some might argue more specifically, of scientism. See, for example, Wieseltier's (3 September 2013) distinction between a *scientific* and *scientistic* standpoint on the humanities.

genetics from social and cultural standpoints[6] have been relatively limited. This may in part be due to the complexities of the science itself and the difficulties that face the non-scientist in gaining critical purchase on communities and practices of which they are not a part. Yet the profound cultural purchase of genetics – a science that has sedimented into popular discourse than most other scientific enterprises – means that it is not, and should not be regarded as, a terrain limited only to scientific practitioners. There are, nevertheless, a distinctive set of critical and intersectional literatures from which this book takes its impetus. Studies about genes and genetics have tended to array into four more or less distinct areas, though of course many of these texts have crossover interests.

First are critical assessments which are focused on the impact of recombinant genetics on citizenship, rights and social inequalities (e.g. Nelkin 1991; Spallone 1992; Tobach and Rosoff 1994; Fox Keller 2000; Ettorre 2002; Thacker 2005; Sunder 2006). These texts tend to be particularly concerned with the social conditions, institutional relations and social consequences of genetics, particularly with respect to questions of gender, race, class, disability and sexuality. These studies examine the impact of genetic practices on rights, on social inclusion, on already marginalised groups, and on the larger terrains of economic, political and social inequality.

A second and linked area of literature is concerned with genetic epistemologies and embodiment. These texts tend to take up poststructuralist and postmodern theoretical approaches and, as with the previous arena, are often strongly located in or influenced by feminist epistemological critical standpoints (e.g. Spanier 1995; Haraway 1997; Steinberg 1997; Katz Rothman 1999, Lemke 2004; Franklin and Roberts 2006). A distinctive subset of this literature is particularly concerned with questions of bioethics, bodily governance and law (Nelkin 1991; Spallone 1992; Shildrick 2005; Rose 2006; Sewell 2009).

Third are texts concerned with media, representation and the gene (e.g. Van Dijck 1998; Turney 1998; Nelkin and Anker 2003; Haran et al. 2007). More recent texts include Nelkin and Lindee's (2004) focus on the iconography of genes and Stacey's 2010 examination of cinematic treatment of the gene. These texts have been concerned with the modes through which genes and genetics have entered the popular imagination and the key representational motifs of genes across a range of contexts, particularly in visual representation.

Fourth are texts that offer speculative interrogations of genetics in terms of scientific capabilities and futures. Books in this cluster, as distinct from books particularly in clusters 1 and 2, tend to articulate powerful investments in

---

6 I am referring here to scholarly rather than journalistic or popularising texts, of which there has been a steady, if also, given the extensive reach of genetic discourse, comparatively modest output.

genetic futures (e.g. Jones 1994; Kitcher 1996; Winston 1997; Woolfson 2000; Frank 2010), but also include more ambivalent and critical assessments (e.g. Rose 2003; Sunder 2006).

Additionally, as Ettorre, Katz Rothman and I suggested in our introduction to *Feminism Confronts the Genome* (2006*)*, the apotheosis of the gene has been 'articulated on a conceptual terrain in which critical ideas concerning reproductive rights, ecology, embodiment, bioethics, choice and agency have been reshaped by feminism.' (p. 134). Of particular note have been the standpoint debates[7] (which challenged conventional epistemological claims about scientific objectivity) and the emergent postmodern turn in feminist science studies to inter-subjective, affective and symbolic relations of science,[8] of which genes are a salient case study.

A Convergence of Cultures

This book, then, aims to enter into a multi-stranded field of debate – informed by an array of intellectual traditions and standpoints – and also to offer particular distinction in a number of ways. Firstly, it aims to elaborate, both substantively and methodologically, on the intersections of science and culture and on the dramatic and pervasive purchase of genes and genetics. It aims to account not only for the social conditions and consequences of genetics, but for their terms of persuasion. It is interested not only in the impact and purchase of genes across social and cultural contexts, but also in their plausibility as knowledge and their power in inchoate terms, as objects of phantasy and attachment. And finally, this book aims also to trace a personal interpretive journey, serendipitous, even accidental, that happened to position me ringside, as a conscious witness of the progressive and exponential genetification of culture.

---

7 For extended discussion of the distinct critical traditions and paradigms within feminist studies of science, see, for example, Kirkup and Smith Keller (1992); Kirkup et al. (2000) and Ettorre, Katz Rothman and Steinberg (2006a).

8 See, for example, the essays in Jordanova's 1986 edited book *Languages of Nature*.

Chapter 1
Languages of Risk: Genetic Encryptions of the Female Body[1]

The new genetics marks the birth of a new kind of medicine. It heralds the most comprehensive assault ever against disease and will have a profound impact on each and every one of us, influencing what we eat and drink, where we live and even the jobs and hobbies we pursue. As children we will know if we will be at high risk in later life from diseases such as cancer and heart disease. As prospective parents we [sic] will no longer face the anguish of a pregnancy that may end in abortion if we are known carriers of defective genes. (Illman, 1 March 1989)

If the biochemical basis for a genetic disease is known, it is usually possible to isolate the *offending* gene. (Weatherall et al. 1986, p. 85, my emphasis)

Britain is Moving Closer to Having "Three Parent" IVF Babies. (Lai, 28 June 2013, *Slate*)

Mitochondrial transfer procedure could prevent mothers passing on devastating genetic conditions to their children. (Sample, 28 June 2013, *Guardian*)

IVF baby born using revolutionary genetic-screening process
Next-generation sequencing could enable IVF clinics to determine the chances of children developing diseases. (Sample, 7 July 2013, *Guardian*)

Introduction

Notions of genetic origins, genetic disease, genetic risk and genetic 'self ownership'[2] have sedimented powerfully into both common vernacular and the fabric of cultural persuasion, if not always into common practice. The

---

1 An earlier version of this chapter appeared under the same title in *Women: A Cultural Review* 7(3), 1996.

2 This coinage appeared in the headline 'Angelina Jolie's Genetic Self-Ownership is the Future of Medicine' (reason.com, 15 May 2013) and reflected a widely validated understanding that Jolie's decision to undertake a prophylactic double mastectomy as a response to her family history and particularly her diagnosis of carrying a BRCA gene.

currency of scientific and popular iconography of the gene – as geographical terrain which can be mapped, traversed and transformed; as text which can be read and (re)written and as 'offending' body which can be 'isolated', altered or eliminated, has come to constitute not only a 'new kind of medicine', as Illman, above, suggested in 1989, but a reconfigured language of reproductive 'health'. This is a language inflected by converging currents concerning bodies and 'risk' – one deriving from gendered discourses of risk, localised to the properties of the female body, and the other deriving from what Lemke (2004) has termed 'genetic governmentality'. Indeed, the advent of *in-vitro* fertilisation (IVF), specifically the capability of producing extra-corporeal embryos, and the increasing proliferation of pre-implantation genetic diagnostic (screening) technologies[3] – both of which are premised on medical intervention on female bodies – have produced languages of genetic risk as a specifically obstetric preoccupation. In turn, and notwithstanding that 'genetic risk' or 'bad genes' can be ascribed to men as well as women, they have conscripted women's bodies as the primary locus of anxiety about 'offending' genes and yet, at the same time, as the absent referent of a genetified reproductive imaginary.

The focus of this chapter is on the place and significance of reproductive medicine as a foundational condition of possibility for the emergence of human genetic science and transformed and genetified understandings of health and illness. In its exploration of the convergence of genetic and obstetric discourses of risk, this chapter makes three arguments. First, it argues that current understandings of genetic risk, driven by but not limited to reproductive medical contexts, derive in part from earlier, historically constituted medical narratives of danger associated with female bodies, specifically of women as dangerous bearers, carers and carriers. Secondly, it argues that understandings of genetic risk are informed by a parallel trajectory of genetic governmentality arising in part from the move from early eugenic preoccupations with human 'fitness' and breeding to contemporary genetic concerns about the care, cure and prevention of illness. Thirdly, it argues that the convergence of these trajectories has constituted a *gendered field* of ethical burden whose regulative and embodied effects are disproportionately borne by women and yet, at the same time, where women are also, typically, displaced subjects and absent referents.

See Steinberg (2014) for further discussion of genetic risk and its emergent calculus of personal responsibility in the context of genetic diagnosis and cancer.

3 Such screening practices include methods of prenatal diagnosis (e.g. amniocentesis), pre-implantation diagnosis (screening of embryos produced by *in vitro* fertilisation) and genetic counselling and screening of adults.

I will begin with a brief contextual discussion of reproductive medicine, and specifically, the advent of *in vitro* fertilisation capabilities which constituted a central precondition of human genetic research and its projective and concrete applications in clinical medicine. I will then briefly set out Thomas Lemke's critical parsing of 'risk' in the context of genetics and his understanding of 'genetic governmentality', particularly his discussion of genetic diagnostics as a 'regime of equity' which both displaces and discursively reconstitutes earlier modes of social-categorical discrimination, but where, interestingly, he does not discuss gender as one of those categories. I will then turn to the central considerations of the chapter: the convergence of gendered and genetified discourses of risk. In this context I will trace a number of thematic trajectories through which modern medical discourse has, from its consolidation in the late nineteenth century, conceptualised female sexuality and sexual-reproductive organs as repositories in which danger, pollution and the transmission of disease and moral degeneracy, in classed and racialised terms, are understood to be immanent. Here, I will consider the constitution of women both as bodies *of* risk and bodies *at* risk, with particular reference to obstetric, psychiatric and eugenic medico-legal discourses in which the regulation of the maternal body and the reproduction of 'fit' families emerge as interlinked logics. I will then turn to the gene and consider the ways in which genetic diagnostic discourse reflects, transforms and reinscribes the female body as primary locus and vector of, at one and the same time, risk, diagnosis and, what Mort (1987) has termed, medico-moral regulation. I will conclude by returning to the question of genetic governmentality and suggest that diagnostic genetics is foundationally a gendered terrain, constituted of asymmetric ethical and embodied imperatives on which its *regimes of equality* depend and at the same time, conceal, and in which the normative reach of 'genetification' and IVF are now mutually expansive.

The IVF Gateway

The advent of IVF in 1978 set in motion both the pragmatic and imaginative preconditions for the development of diagnostic genetics. Because IVF procedures produced extra-corporeal embryos, it became possible to undertake myriad forms of research on human genetics and also to develop processes through which embryos could be diagnostically screened before being implanted back into a woman's body. According to the records of the Human Fertilisation and Embryology Authority (UK), the first baby born to an IVF patient following the use of pre-implantation genetic diagnosis was in 1990, at the Hammersmith Hospital.[4] IVF was thus both a gateway technology and a normalising context

---

4 Steinberg (1997), 105. See also Winston (1997; 1999).

for the development of diagnostic genetics and for its wider discursive impact on the popular and clinical-biological imaginary. At the time of finalising this book, pre-implantation diagnostic capacities are well established and normative aspects of IVF practice and the genetified understanding of risk, health and illness has fully embedded into common vernacular.

The advent of IVF was also significant in another way, far less generally acknowledged and far more taken for granted; it involves a range of invasive interventions that can only be practiced on women. IVF and related interventions have, in this way, followed the character of reproductive medicine which has historically skewed towards female reproductive processes (including where the reproductive problems being addressed arise from male bodies).[5] IVF procedures aim to (surgically) displace the locus of conception from the female body in order to create disembodied embryos. Typically this involves an at the least uncomfortable[6] and usually repeated regime of hormone drugs and laparascopic extractions undergone by women, with fairly low odds of a viable pregnancy and live birth at the end.[7] The gender specificity of IVF treatment generally goes without saying. And yet it is a critical precondition for the genetics enterprise. Both tendencies are captured by the citations at the start of this chapter, which span IVF based genetic diagnostic practices from 1989 to 2013.

Genetic Governmentality

Thomas Lemke (2004) provides an interesting and useful parsing of the transformed constitution of 'risk' in the wake of genetics, and one that is particularly apposite for the focus of this chapter. He argues firstly that genetic 'risks' are effects of 'problematisation' (ibid., 552) as distinct from 'biological-empirical matters of fact'. This promotes an individualised imperative of prevention (i.e. definitive, certain action) premised on a calculus of maybes. Genetic diagnosis, even where it appears most empirically solid, nevertheless profiles statistical, population based correlations and possibilities; it does not and cannot predict individual outcomes. Genetic testing thus constitutes a

---

5 For more extended discussion of the use of IVF (on women) to treat male infertility, see Steinberg (1997).

6 For some women, the procedure can be extremely painful and high risk as well as disruptive and uncomfortable (Steinberg 1997; Throsby 2004).

7 A recent *Guardian* article (Sample, 17 May 2013) suggested that IVF pregnancy rates may be on the cusp of significant improvement. It is additionally important to note that IVF procedures are not limited to IVF treatment as they are also used to extract ova for 'egg donation', where the ova may not be fertilised or reimplanted.

totemically charged field for the containment of uncertainty – a plan of action, a concrete intervention against the spectre of possibility. Second, Lemke notes that genomic research occupies a terrain of complexity (which it is increasingly acknowledging)[8] but is nevertheless characterised by an epistemically reductionist orientation. This spans the 'canonical human' of the human genome project – the *consensus genome* as a 'uniform genetic standard' rather than a reflection of any individual person's DNA (p. 553) – to the investigation of the relationship between single (or even clusters) of genes and disease. Thirdly, Lemke argues that genetic research constitutes 'a discourse of deficiency'[9] which analyses life by means of concepts such as 'absences', 'faults' and 'defects' (p. 554). At the same time, and in so doing, genetic diagnostics constitutes a *'regime of equality'* (p. 557); genetic 'flaws' can afflict anyone. One effect of this regime, Lemke suggests, is that it shifts the terrain of inequality from the social-categorical (such as race, ethnicity, gender, class), to the naturalised and individuated; inequality arises from the nature of the 'flawed' individual (ibid.). Another effect, I would suggest, is the concomitant naturalisation and negation of the subject position of women *vis-à-vis* the genetics enterprise. Gender inequity, in other words, is central to genetics' regime of equality.

Languages of Risk 1: The Female Body

Many commentators have identified the nineteenth century as an era which not only established science and medicine as dominant professional discourses, but as a period in which medicine and science became invested with moral regulatory power. Usher (1991) and Mort (1987) among others,[10] have also, in various contexts, identified the Victorian consolidation of professional medicine as a process that linked both disease and biological reproduction, specifically female reproductive processes, with racialised, class specific and heteronormative understandings of physical, social and moral danger. This was the era when race was constituted as a scientific category and when race, class, gender and other social divisions idealised in Victorian Domestic Ideology became rationalised

---

8 There has been a move from single genes to genomics (the genetics of a whole organism), for example, and also more contextual understandings, as constituted in the more recently emergent field of epigenetics (the intra and extra molecular effects on the expressions of genes) and microbiomics (the interactive epigenetics of the multiplicity of organisms that constitute a body). However these coexist with rather than displace the gene focused reductionism discussed by Lemke and others.

9 This phrase, Lemke takes from Gottweiss (1997).

10 See, for example, Oakley (1976); Nead (1988); Davis (1990).

through the lens of biology.[11] The consequent rise of the Victorian eugenics movement produced new discourses of legitimate and 'illegitimate' family, mediated through languages of pathology, and new languages of morality, social order and law mediated through the professionalisation of medicine. This was also an era where the construction of the female body as both an *embodiment* of and a *carrier* of danger and risk, was reconstituted as biological, as opposed to demonological, fact.[12] In all these contexts, the female subject emerged, in the Foucauldian sense, as *docile* (disciplined and self-disciplining), *divided/divisible*, and in both subjective and objectified senses, as *bodies of resistance*.[13] Indeed, it can be argued that the medicalised reconstitution of the risk discourses of the female body and female subjectivity was central to the symbolic, familial and imperial[14] economies of the Victorian era.

Dangerous/Docile Bodies

I am not the first to note the preoccupation of modern medicine with decoding the mysteries of female bodies and managing the dangers or dysfunctional potentials which are seen to inhere in female reproductive processes and female agency.[15] As Oakley (1976) suggested in her seminal early work, the formation of Obstetrics as a medical specialism (and indeed of medicine more generally as a masculinised profession) was constituted, in no small part, through an association of women with danger at a number of levels: as bearers and custodians of knowledge (dangerous healers), as mothers and as maternal

---

11 Proctor (1988) has argued that one of the key functions of science , including biology and medicine, is *apology*, that is, the naturalisation of social processes, specifically of social inequalities. See also Roberts (1992) for discussion of the evolution of race and gender ideologies in the context of the USA.

12 See, for example, Oakley (1976); Witz (1992). It should be noted that Witz critiques Oakley's 'strong thesis' of the medicalisation of medieval demonological constructions of femaleness as overly simplistic. However, she notes that modern medical discourses of the female body (and the changing market relations of medical practice) cannot be adequately understood without reference to the complex history of religious discourse surrounding female danger dating back to the fourteenth century.

13 Thus, I am not arguing that women were/are constituted as passive objects of discourse, but were/are interpellated (produced by and produce themselves within and in resistance to) dominant medical constructions of female bodies.

14 The constitution of the female body as metaphoric object of sexual/imperial conquest (and of the colonised land as female body) were for example, central to the literary idealisation of British Empire (see, for example, Bristow (1991); Stott Summer (1989)). See also Roberts (1992) for a parallel discussion of the interlocking of race and gender from slavery through eugenic ideologies in the US context.

15 See, for example, Corea (1979); Douglas (1966); Lupton (1994).

body. In this context, both female controlled reproductive care and the physical reproductive bodies of women were invested with notions of risk, danger and disease.[16]

The masculinisation of healing generally, and of obstetrics specifically, not only underscored a significant loss of female autonomy over the female body, but placed women at risk from new dangers of invasive (very often surgical) medical management of pregnancy and childbirth.[17] An increasingly *recombinant*[18] logic can be traced in this process. The investiture of medicine as a profession was grounded, for example, on a middle class, masculine monopoly of surgical tools and techniques – a development in turn grounded in a discourse, characteristic of Western allopathic medicine, of bodies as composite, component, re-arrangeable parts. Both Usher (1991) and Witz (1992) argue that the surgical management of female bodies was central to the economic viability and closure of professional specialisms including obstetrics and psychiatry. Usher has suggested that if medicalisation underpinned the stigmatisation of women as dangerous bearers, carers and carriers (of physical and moral disease), psychiatry singled out the organs and properties of the female body as harbingers of madness and generational (i.e. inherited) 'degeneracy'. Both Usher and Witz furthermore note the role of surgery on female reproductive–sexual organs[19] as the central avenue through which were consolidated both the gendered and class closure of the medical professional and the construction of female bodies and subjects as dangerous breeders of individual and social chaos. Both obstetrics and psychiatry were, in other words, disciplinary regimes through which risk became understood as a primary and normative property of female bodies and female subjectivity.

---

16 See also, Usher (1991); Witz (1992). It is worth noting that Witz is critical of Oakley's 'strong thesis' that an allied patriarchal triumvirate of law, medicine and religion conspired to disenfranchise female healing and medicalise/criminalise the female body. Witz suggests that a more complex and contestatory power struggle took place among these forces and that the rise of capitalism and Victorian class struggles were among the key constituent conditions that produced a masculinised profession of medicine with a much later rise of obstetrics and gynaecology as a medical specialism.

17 Indeed, Victorian medicine contributed significantly to maternal mortality (Oakley 1976).

18 A *recombinant* logic can, more broadly, be said to characterise the rise of industrial capitalism, epitomised in the assembly line production process which produced both labourers and products as component rearrangeable parts.

19 These included surgical management of birthing in the context of Obstetric medicine and treatments involving the amputation of sexual/reproductive organs (clitoradectomy, ovariectomy, hysterectomy) in the context of Psychiatry.

Bodies of Contagion

The growing currency of obstetric ideas of the female body also mapped into a complex matrix of anxieties about sexual morality[20] and familial and reproductive 'fitness'. In this context, medicalised languages of risk emerged as the vernacular of class and racial as well as sexual difference. For example, the interface of emergent germ theory (of the transmission of disease) with the 'war on dirt', characteristic of Victorian social purity campaigns linked imperatives to 're-educate' the working class, immigrants and non-white/non-Europeans[21] with anxieties about breeding. Concerns about 'race suicide' and miscegenation (interbreeding across racial/class divisions) specifically emerged from medicalised discourses that conceptualised dirt, germs and disease as *embodied* by and emergent (as 'miasma') from these stigmatised groups.[22] These ideas underpinned a growing medico-legal apparatus of reproductive and sexual regulation, including compulsory sterilisation, miscegenation and anti-abortion laws, which were particularly directed at the female body.[23]

The passage of the Contagious Diseases Acts (CDAs) of the late nineteenth century in Britain and British colonies provides a stark illustration of the consolidation of eugenic and obstetric discourses of danger. It is in this context that the female body, specifically the body of the working class prostitute, was singled out and stigmatised as a transmission vector for both sexually transmitted disease and class corruption. Ostensibly designed to contain the spread of sexually transmitted diseases, the CDAs established an apparatus of mandatory registration and medical examination of women working as prostitutes. As noted by Walkowitz (1980) Mort (1987) and Nead (1988), the fact that sexual surveillance was not directed at men meant that not only were the Acts totally ineffective at controlling the spread of syphilis and other illnesses, but women's bodies became the symbolic and literal terrain of the class inflected war on disease, dirt and immorality. In this context, the body of the prostitute became emblematic of social chaos, the threat of socialism, the destruction of the sacredness of marriage and 'good' family and of the monstrous sexual and reproductive congress of 'respectable' and 'unrespectable', 'fit' and 'unfit', classes.

---

20 See Walkowitz (1980); Mort (1987).

21 See, for example, Burton (1990); Nair (1992) for discussion of British imperialist 're-education' of Indian women.

22 See, for example, Gilman (1991); Mort (1987); Proctor (1988); Solomos (1989) for discussion of the complexities of particular groups interpellated into pre-World War II European and American eugenic discourses.

23 See, for example, Davis (1990); Koonz (1987); Steinberg (1996b).

Thus, the CDAs represented clearly the interplay of eugenic and obstetric languages of risk. Here it was specifically the maternal-sexual body that invested with the class inflected and racialised pollution beliefs characteristic of both racial hygiene and social purity ideology. Indeed, the conjunction of medical and social purity ideology produced a language of contagion in which class, race and gender inflected notions of germs, generational inheritance (genes), disease and immorality were dynamically interlinked. In this context, the female body not only emerged ideologically as site and source of contagion and danger, but also materially as primary target of and transmission vector for sexual and reproductive regulation.

Divided Bodies: The Antagonistic Pregnancy

The languages of risk and recombinant, eugenic logics that underpinned emergent obstetric constructions of women and female bodies also underscored the incipient medico-legal construction of pregnant women as divided, adversarial bodies.[24] The contemporary construction of a pregnant woman as not only two people, but two patients with separate and, indeed, hostile interests has its roots in the transformed conceptualisations of the maternal body, as dangerous vessel, from the early nineteenth century. As Fyfe (1991) notes, 1803 saw the first British law that criminalised abortion, thus extending notions of maternal danger to the female will.[25] Many commentators[26] have traced the growing embryo-logic orientation of obstetric technologies and increasingly surgical/invasive medical treatment of birthing women with their consequent erosion of women's self determination over the management of their pregnancies and the conditions of their birthing. This is perhaps epitomised in the modern-day imagery (and reality) of the pregnant woman strapped down and connected externally and internally with 'fetal monitors' and in the increasing number of cases where hospitals have sought and won legal power to suspend women's right to refuse treatment, such as in court ordered caesarean sections, for the 'welfare' of the fetus.[27] The vernacular of 'fetal surgery' encapsulates this logic in its linguistic obfuscation of the fact that such surgery is performed on pregnant women. Franklin (1991) has traced the underpinning construction of women as dangerous docile bodies housing a fetal martyr-victim-hero in the 'fetal fascinations' (p. 205) of embryology and fetology – specialisations

---

24 For extended discussion of this point see Steinberg (1991; 2000).

25 Fyfe (1991) points out that 'this law criminalized, for the first time, those who have "intent" to procure abortion' (p. 162).

26 See, for example, Corea (1979); Spallone (1992); Reissman (1992); Lupton 1994).

27 See, for example, Feinman (1992); Merrick and Blank (1993); Colker (1994); Dubow (2010).

that have emerged out of the specialisms of obstetrics and gynaecology. The pervasive construction of maternal danger, which underpins the ethos of both anti-abortion and fetological discourses, also constitutes a key conceptual trajectory through which has emerged the putative 'need' for embryo protectors who by definition cannot be (pregnant) women and the ongoing medico-legal-religious struggle for 'protective' custody of the fetal territory of pregnant women. Along side these discursive, medico-legal trajectories has been the progressive invention and use of technologies aimed to image, manage and, diagnostically screen the maternal-pregnant body, with its ambivalent effects of empowerment and disempowerment of pregnant women's choice and rights.[28] In this context, the language of 'genetic risk' cannot but draw from, even as it may obscure, the many stranded and historically constituted languages of risk that have long been associated with the female body.

Languages of Risk 2: Gender and the Gene

Diagnostic genetics refers to the identification of disease inducing, or disease potential, characteristics understood to be located in the molecular constitution of the body. Genes, like germs, are associated with definitions of disease that encompass both physical ailments and conditions as well as undesired social characteristics. Thus we are said to have genes for Tay-Sachs and gay genes, genes for haemophilia and genes for criminality, genes for schizophrenia and genes for sensitivity to toxic chemicals. The progressive ascription of an ever widening array of conditions to genes as sole or partial determinants represents a genetified expansion of the diagnostic field and its associated medico-morality – what Lemke (2004), among others, has referred to as 'genetic imperialism'. Moreover, gene theory reifies that current within germ theory which locates the origins of disease as a defect within the body, family line, character and social profile of the individual.[29] Lemke describes this as the 'privatisation of social risk' (555). Thus, as with diagnostic medicine historically, genetic diagnosis can be said to carry social and moral resonances and to mediate, whether implicitly or explicitly, social-categorical notions of reproductive heritage, familial 'fitness' and health.

The progressive expansion of prenatal and, following the advent of IVF, pre-implantation genetic screening technologies, reflects and invests in the powerful currency of genetic perspectives of health and illness. As suggested

---

28 See, for example, Rothman (1987); Petchesky (1987); Spallone (1989); Merrick and Blank (1993); Steinberg (2000).

29 For extended discussion of genetic reductionism, see Hubbard and Wald (1993); Spallone (1992); Steinberg (1996a); Lemke (2004).

by Illman's early prognostication, quoted at the start of this chapter, the logic of genetic screening carries an incipient potential to interpolate not only reproductive life, but indeed, all aspects of our lives. His construction of genetics as an 'assault' on disease is significant not only for its recapitulation of military metaphors typically associated with medicine,[30] but for the particular ways in which putative genetic 'enemies' elide 'defective' genes and 'carriers'. The power of such a discourse to intensify the stigmatisation of illness and disability should not be underestimated. As Sue Meek (Meek 1986), another early theorist of emergent genetic diagnostics, argued:

> Knowledge of the result of genetic screening tests may adversely affect marriage and employment prospects and health insurance status...The development of increasing number [sic] of these tests is likely to exert a pressure for routine genetic screening of the foetus itself rather than just members of identified, at risk, family groups. Peer group denigration and potential prosecution by a child born with a genetic defect may await parents who decide not to have such screening performed, or who decide to continue with the pregnancy despite the outcome of the tests. (p. 41)

'Defective' Bodies

The language of genetic risk characterises the conditions associated with 'bad' genes as both an inevitable tragedy and a natural disaster, twin hallmarks of what Lemke (2004) calls a 'discourse of deficiency'.[31] Thus, in a genetified frame, negative social attitudes towards illness and disability emerge as consequent to the condition itself, rather than, for example, material social conditions that disempower and disable. Thus genetic screening has a preventive orientation directed, not at the power relations and social structures surrounding disability, but rather at controlling the reproduction of (those bearing) 'bad' genes. In this sense, the genetic diagnostic gaze would appear to contribute to what Illich (1976) termed 'cultural iatrogenesis', that is, the process of attrition through pervasive medicalisation, of the belief that one can assimilate and cope with pain and illness. The tendency of genetic discourse would, thus, seem to foreclose on a social politics of disability. Genetic screening also constitutes a reinvestment in 'perfectly healthy' offspring and 'normal' family,

---

30 Sontag's *Illness as Metaphor* (1991 [1978]) examined the military metaphors associated with illness, an analysis that has informed sociological, cultural and feminist studies of medicine more broadly (see Lupton 1994).

31 Elsewhere (Steinberg 1996a) I have explored in detail the reconstitution of illness and disability through the lens of diagnostic genetics. See also Hubbard and Wald (1993); Rosoff (1994).

notions which cannot be divorced from the stigmatising resonances that have historically characterised conventional family discourse. Indeed, it can be argued that the disabled body has become the new and acceptable medico-moral cipher for reproductive danger. This is a reconstructed eugenics whose language of health and ostensibly non-coercive character (no one is forced by law to undergo genetic screening) would seem to obviate, or at least be at some distance from, the explicitly racialised, classed and gendered preoccupations of earlier selective (eugenic) rationalities. Yet what does choice mean in the context of a language of 'risk'? And given the disturbing parallels between historical and contemporary 'risk groups'[32] and the pathogenic character implicitly (even if not intentionally) ascribed to bearers of 'bad' genes, it is clear that genetic discourses of risk resonate with and reproduce the heritage of pollution beliefs which they might seem, as many geneticists have argued,[33] to disclaim.

'Defective Carriers': Selective Burdens and the Female Body

There are a number of assumptions about the position of women and the female body in the genetic screening context, assumptions which are obscured in the gender 'neutral' language of 'couple', 'parent', 'individual' and, indeed, 'offending gene'. However, if we consider the procedural context and character of genetic screening, its gendered resonances become clear. While, as I noted above, genetic 'disease' or 'bad' genes are not seen as the exclusive property or properties of the female body, it is women nonetheless who bear the primary medico-moral burdens – procedural, conceptual and in terms of choice – of both genetic diagnosis and notions of genetic risk.

Firstly, obstetrics is a central context for the development and practice of genetic diagnostic techniques. Thus, it is the medical specialisation concerned with the pathogenic potentials within or affecting the maternal body which serves as a primary arena for genetic screening. From pre-implantation to prenatal diagnosis practices, women's bodies are primary vectors through which

---

32 Gay identity, 'criminality', mental illness, 'low' intelligence, are some of the 'conditions' for which proof of genetic origins are being sought. Additionally, many key disease categories continue to be attributed to and elided with various racialised/ ethnic groups; thus Greek populations are seen to produce Thalassaemia and have Thalassaemia genes; Ashkenazi Jews to produce Tay-Sachs, have Tay-Sachs genes etc. Within the diagnostic genetic gaze, ethnic, racialised and class origins as well as sexual identities have a narrowly diagnostic significance and become (re)enscribed as both pathological and pathogenic.

33 See Steinberg (1993) for a discussion of disclaimers as they appear throughout the texts early advocates authored by medical scientific practitioners who advocate the development of genetic screening.

assessments of 'risks' are investigated (and which may, on this basis, be further subject to therapeutic imperatives of intervention post-implantation). Thus the female body emerges as the primary site of anxiety about genetic risks, whether these are seen to accrue from her own genetic heritage or not. In turn, and as a consequence, it is the female body which bears the material burden of risk before and during pregnancy. With respect to prenatal diagnostic techniques, these may include the risk of miscarriage and the physical and emotional consequences of terminating a wanted pregnancy. With pre-implantation diagnosis, there is the necessity of first undergoing IVF, itself a risk laden process involving hormone treatment (to induce ovulation) and surgical interventions (for removal of women's eggs).[34] Thus while women's bodies are in several senses *taken* as present, they are also discursively *positioned* as absent. This is reinforced through the ostensibly egalitarian, ungendered languages of 'couples', 'parents' and 'genetic risk' while the particular burdens and risks of genetic screening to women are discursively obscured.

Barbara Katz Rothman (1987), in her early seminal work examining the implications for women of prenatal diagnostic screening, raised a second set of issues concerning the character of reproductive choice in this context. As Rothman's interviews with women suggested, the investment of pregnancy with the question of genetic risk rendered woman's relationships with their own pregnancies as tentative and fragmented (the pregnancy could not be part of them until 'abnormalities' were ruled out by the tests).[35] This alienation was fostered both through the embryo-centredness of concerns in the obstetric diagnostic context and the concomitant consideration of women as risky 'maternal environments'.[36]

Rothman argued, furthermore, that genetic diagnosis involves a process of 'evaluat[ing] disabilities, deciding which disabilities make life not worth living for the disabled person, and also which disabilities demand too much or are beyond [mothers'/parents'] competence' (ibid., p. 160). Genetic screening requires the making of eugenically framed decisions and then living with

---

34 For extensive discussion of the IVF procedure and health risks involved (including a review of the medical scientific literature on these) see, for example, Klein and Rowland (1988); Spallone (1989); Steinberg (1996a); Throsby (2004).

35 Hundt et al. (2008) found a collateral effect in their more recent research on women's experiences of antenatal screening, where women became invested in a positive diagnostic finding, possibly in part because diagnosis signalled an entitlement to care. Paradoxically, this did not necessarily produce further alienation, but rather also could enable reconnection with their pregnancies.

36 See also Petchesky (1987); Spallone (1989); Science and Technology Subgroup (1991); Feinman (1992); Merrick and Blank (1993) Gregory (2012) for further discussions of prenatal imaging and broader contextual questions concerning adversarial constructions of pregnancy.

the consequences, a decision-making process which Rothman's interviewees described as anguishing, raising the question of distinguishing choice from *desirable* choice (pp. 181–92). While men might share in the decision-making, the fallout of choices based on the information provided by genetic diagnosis will nevertheless be disproportionately borne by women; it is women who will bodily live out that decision, whether it means a continuation or termination of a 'risky' pregnancy. That it is women who are primarily held responsible for and characteristically undertake primary care of children can only increase the pressures on women's reproductive choices in the screening context. The pressures to (re)produce 'healthy' children, extend (and extend from) assumptions that women are primarily and normatively responsible for social-material labours of family. They are also intensified in a context where disability is profoundly stigmatised and where material support for carers and people living with disabilities are pervasively lacking. Thus languages of 'genetic risk' reflect, interlink and reproduce the oppressive relations surrounding both disability and motherhood.

Moreover, as with earlier pollution beliefs, the notion of 'genetic risk' has particular resonance and consequences for women as bearers, carers and carriers. Rothman argues that genetic screening is informed by the construction of women's bodies, reproductive process and wills as 'if not the enemy, then the battleground' (p. 111). This adversarial construction of the female body is heavily leveraged by the adversariality that pervades the 'culture wars' surrounding abortion and the larger questions of women's bodily sovereignty, reproductive rights and (de)legitimacy as citizens and social actors. In this context, the 'guilty' burden of genetic risk is, I would suggest, disproportionately weighted onto the female body, whose reproductive processes are already viewed with concern, if not opprobrium within dominant obstetric and legal[37] discourse.

Indeed, against (and as an extension of) historical configurations of risky female bodies, it can be argued that women become implicitly construed as bearers of 'nature's defects'. The putatively pathogenic potential of the maternal body maps almost seamlessly into its potential as 'gene transmitter'. The maternal body emerges as the primary vector for the transmission of 'bad' genes whether the origins of those genes are male or female. In this sense, the female body becomes the dangerous carrier of dangerous carriers. The 'bad' seed of men, the 'bad gene' itself, the 'defective' embryo and the 'damaged' generational heritage become incorporated materially and symbolically within the maternal body.

Finally, insofar as genetic screening involves the policing of the reproduction of 'abnormality', it resonates with broader conceptualisations of reproductive 'fitness', even as these are more likely to be deployed in terms of 'health'. Notions

---

37 And, in much of the world, in religious discourse.

of normal and abnormal familial heritage, characteristic of conventional family discourse frame and become reconstituted as genetic mechanisms which can be, apropos current parlance, 'switched' on and off. In this context, 'offensive' genes and the diagnostic orientation in and of itself invoke, even as they appear to replace (and displace) earlier, more overtly racialised and classed discourses of 'fitness'. Genetic diagnostics is in this sense a genealogical enterprise that does not, as Lemke suggests, trade off disciplining or discrimination for design (Lemke 2004, p. 562), but rather premises design on discipline and discrimination.

Conclusion

If we return to the quotes that began this chapter, we can see evidenced concretely the ways in which they have become enmeshed in popular representation. Firstly, they point up tensions surrounding reproductive agency of women *vis-à-vis* medical discourse. Clearly the constitution of the female body in terms of danger frames and delimits both the scope and character of reproductive choices women make, including those which are aimed to resist the controlling trajectory of obstetric, eugenic discourse. In this context women face a particular and distinctive paradox. The promises of freedom from disability (and its stigmatisation) and from the disempowering and often punitive social consequences of bearing and caring for a disabled child, both of which are implicit in genetic diagnostic screening, are clearly potent. They are powerful not least because of the social expectations and empirical realities that women have primary responsibility for the caring labours of parenting. At the same time, the choice to undertake genetic screening can only reinvest in the political field that make screening seem necessary.

Second are the mutually normalising and mutually expansive effects of IVF and genetic diagnosis. In 2013, both genetic pre-implantation screening and IVF practices are invoked as fully established and unremarkable features of both the obstetric landscape and the popular imaginary. Moreover, where IVF began in 1989 as the gateway technology for the development of human genetics, the expansion of genetic diagnostics has become, reciprocally, a gateway and powerful rationale for the expanded use of IVF. As genetic screening becomes more elaborated, its imperatives of action cannot but add to the pressure to 'choose' IVF in order to be 'safe', to be 'responsible'. That these expansive tendencies must concomitantly conscript, and particularly rely on, female bodies and labour is both a primary condition and a point of negation. In the genetified field of risk, gender is imperative and disturbingly absent. The taken-for-grantedness of this asymmetry is perhaps more striking given the extraordinary pitch of controversy surrounding the (attacks on) women's

23

reproductive health and rights, particularly in the USA, that have regularly made the news in 2013.

It could be argued that the particular burdens placed on women with the genetification of risk would seem an inevitable consequence of the fact that pregnancy takes place in and of women. Thus a material 'democratisation' of genetic diagnostics (as distinct from the symbolic 'equity' of its language and representation) would seem to be precluded by the empirical realities of biology as well as the facts of technology. Nor does it seem particularly empowering for women if, and as, men are increasingly interpellated into the 'guilty' burden of genetic risk. Indeed such a question would seem to entrench the naturalisation of 'risky', 'unfit' and 'defective' bodies as individuated products of biology. Against the historical trajectories of obstetric and eugenic languages of risk, therefore, it is hard to imagine how genetic screening could avoid, or more importantly, challenge the reification of the female body in particular, and in all the terms explored above, as vector of danger, diagnosis and medico-moral discipline.

Chapter 2
Metaphor, Materiality and the Gene[1]

Languages of nature are made and not pre-given (Jordanova 1986, p. 27).

Introduction

In her classic essay, *Illness as Metaphor* (1977/1991), Susan Sontag examined the symbolic economies of disease in order to trace and yet disaggregate the inter-resonances of medical scientific discourse and commonsense. In the dominant metaphors associated with illness – including, for example, military, criminological, capitalist and machine metaphors – Sontag intimately mapped what she identified as the conceptual and material preoccupations of modernity. In this context, language, and metaphor specifically, mediate historically contingent and reciprocal investments between medical scientific discourse and broader cultural processes and meanings. In this reading, *Illness as Metaphor* not only articulates the cultural *situatedness* of science and medicine but implicitly posits them as sites of both cultural agency and cultural authorship.

Yet, one of the peculiar contradictions of Sontag's early study is the contrast posited between metaphoric thinking as a purely linguistic, extra-scientific and secondary phenomenon, and the scientific standpoint. Sontag was concerned with the effects of metaphoric language, its capacity to distort fact and reality and to subvert scientific inquiry. Indeed, Sontag argues strenuously for the exposure of destructive metaphoric representations of illness as part of a process of conceptual decontamination. This positivistic investment in the putatively mimetic and objective qualities of medical scientific standpoint strikes a peculiar note in several respects. For example, while Sontag tracks the political implications of particular kinds of metaphors associated with certain diseases (and disease *per se*), she understand these to reflect a cultural overlay. This construction would seem both to reject the ways in which metaphoric thinking demonstrably fuels scientific research (and the scientific imagination) and the ways in which science *resources* the broader cultural take-up of particular disease meanings. Ironically, many of the sources Sontag quotes in *Illness as Metaphor* were those of medical scientists. In this reading, metaphor itself

---

1 An earlier version of this chapter appeared in S. Williams et al. (eds). 2000. *Theorising Medicine, Health and Society*. London: Routledge.

becomes metaphor – of the collision of culture and knowledge and the consequent derationalisation of both.

The positivist and anti-positivist tensions of *Illness as Metaphor* take on a particular poignancy at this contemporary historical moment. On the one hand, nearly four decades on from Sontag's original writing, the notions of science *as* culture, and of the contingent and historically specific relationships between ontology and epistemology, knowledge and power, 'fact' and 'artefact'[2] have complexly reconstituted the epistemic centre of social theory across disciplines. At the same time, resurgent biological explanations of identity, social patterns and social relationships, have harnessed new currency in the iconic configurations of the gene and the neo-positivism of recombinant genetic science. It is this clash (and mutual interpellation) of cultures, as it were, with which this chapter is, in part, concerned.

Narrative, Metaphor and the Gene

> Biology is inherently historical, and its form of discourse is inherently narrative. Biology as a way of knowing the world is akin to Romantic literature, with its discourse about organic form and function. Biology is the fiction appropriate to objects called organisms; biology fashions the facts "discovered" from organic beings. Organisms perform for the biologist, who transforms that performance into a truth attested by disciplined experience; i.e., into a fact, the jointly accomplished deed or feat of the scientist and the organism. Romanticism passes into realism, and realism into naturalism, genius into progress, insight into fact. *Both* the scientist and the organism are actors in a story-telling practice. (Haraway 1989, p. 5)

There can be no doubt that the gene, that modern artefact of nearly two centuries of scientific preoccupation with the 'discovery' of life's origins and the 'mechanisms' of its heredity, has become sedimented into the fabric of contemporary popular imagination. We are now well into the fifth decade of recombinant genetics, which has fuelled not only a biotechnology revolution in its own right, but has imbricated and articulated with the digital media's radical transformation of the knowledge ecology. Genetics has constituted not only novel molecular interjections, but a process of grand narrative. In this context, the gene has become at once a point of reference and of projective

---

2 Haraway (1989) argues that 'fiction is inescapably implicated in a dialectic of the true (natural) and the counterfeit (artifactual) ... [both] rooted in an epistemology that appeals to experience' (p. 4) and linked through processes of scientific and extra-scientific story-telling.

identification, (re)constituted as both a tool and a symbol, taken up in both literary and literal senses as a rogue or domesticated body, a territory, a weapon, a library, a laboratory. The iconicity of the gene both *in* and *as* contemporary commonsense accrues from a complex authorship in which the authoritative accounts of scientific practitioners intersect with the wider array of cultural referents. Indeed, emergent discourses of the gene have complexly and contradictorily filtered through the range of representational media (literary, press, digital, televisual and cinematic, professional and lay) and genres (technical, peer review, documentary, science fiction, police procedural, horror). By the same token, as this chapter will argue, gene stories have conscripted and transformed dominant cultural narratives, from origin to adventure stories, from the dystopian to the utopian.

Drawing, then, from a critical engagement with Sontag's original study, this chapter will examine imagery, metaphors and narratives that underscore the cultural primacy of the gene. In so doing, I am not suggesting that the narrational dimensions of science are in themselves problematic. Rather, I would argue that what is problematic is a *denial* of science as a site of narrative. What is of critical importance then, are the *particular* narratives embedded in a *particular* set of scientific practices and what they suggest about the power relations accruing to that science. Focusing on a selection of what might be classified as 'medium to high-brow' popular science texts as a case-study, I am interested here in the representational economies surrounding genes, bodies and embodiment in three respects: the constitution of genes as bodies/embodiments, of genetics as a body of knowledge and of geneticists as a body of knowers.

Science, Narration and Embodiment: A Critical Standpoint

In reading the metaphors and narratives of genes, this chapter is informed by two particular epistemological trajectories within social theory of science, medicine and the body. Firstly, there is the extensive interdisciplinary literature on the body and on cultural praxes of embodiment. Such work provides a key theoretical resource for this chapter. As I have noted elsewhere (Steinberg 1996a), the body as *corporealisation* of social and cultural relations has become an increasingly important site of inquiry across a range of (inter)disciplines as well as a field of study in its own right.[3] There has been considerable analytic preoccupation with bodily boundaries – fragile, transgressed, defended,

---

3 These include: the sociology of health and illness and of medicine, gender and sexuality studies (of health, medicine, science and the body), cultural studies, bioethics, biopolitics and biotechnology.

reproduced[4] – with embodied spaces from molecular to cosmic, and with the mapping and remaking of bodily function and contour. This emergent interest in bodies and embodiment constitutes, in part, an attempt to reconcile materialist and postmodern standpoints within social theory – to re-incorporate bodies in the matrices of imagination, memory and meaning *without* either collapsing or denying the relationship between being and knowing. A central strand of embodiment theory is specifically invested in the analysis of medicine and science as key discourses and disciplinary regimes through which bodies – bodies that matter (Butler 1993) and bodies of knowledge (Foucault 1977) – are constituted.

Secondly, in taking up the method, but not the positivistic investments of Sontag's original study, this chapter articulates with broader interests in the narrative processes of and within science – and of science itself as a story-telling enterprise. As suggested in Sontag's work, narrative analysis foregrounds discursive relations of authorship and readership and the textual praxes that constitute the languages, protocols and products of science and medicine. In doing so, the complex elaboration of science as product of, and as productive of, wider social and cultural relations can be intimately mapped. In turn, the consideration of science as a project of narrative is part of a larger interest in narrative as a cultural process constitutive of competing concepts of kinship,[5] of imagined community and nation,[6] of identity[7] and of the body. As Redman (1997) has argued:

> [N]arrative is a significant mode through which meaning is produced and lived out: in other words … narrative is important to the ways in which we understand the world and our place in it. This is a fundamentally anti-positivist view point, and it is therefore unsurprising that much of this work has also stressed the narrative construction of knowledge. (p. 4)

The association of narrative with science and, more to the point, the investigation of science as a narrative practice, radically reinterprets its epistemic character and powerfully challenges its capital as a privileged (meta)cultural discourse. In such an analysis, moreover, the boundaries of both discipline and genre that are conventionally understood to demarcate expert and popular knowledges

---

4 See, for example, Bordo (1993); Martin (1994); Butler (1993); Grosz (1994); Waldby (1996); Shildrick (1997).

5 See, for example, Strathairn (1992); Franklin et al. (1999); Franklin and Roberts (2006); Franklin (2008).

6 Anderson (1983); Yuvall-Davis (1993); Brah and Coombs (2000).

7 See, for example,, Dawson (1994); Redman (1997).

are blurred while the underlying inequalities that produce their margins and centres can be starkly revealed. Narrative, moreover, as a site of hegemonic struggle, is deeply intertwined with the valedictory claims of expert tellers and the discursive practices that produce both bodies of knowledge and bodies who know. Thus, a narrative approach to the analysis of medical scientific discourse cannot but reveal its interpretive preoccupation with bodies – the very stuff of 'nature', the very bedrock – object and agentic – of scientific materialism.

Genes and Genre

> [T]hey [scientists] map for us and for themselves the chains of associations that make up their sociologics. (Latour 1987, p. 202)

Authoritative accounts of scientific practitioners cannot be easily disentangled from their renderings for popular consumption. Indeed, the multiple genres of scientific writing and writing (or other representations) about science seem to merge at various points of nexus. The scientist who writes for peer review may also, for example, write a novel, a newspaper editorial, a film script or a handbook for the lay reader, consumer or patient.[8] As media personalities or speakers on the high-brow lecture circuit, scientists characteristically produce themselves as veritable prophets of modernity: translators and transposers, magicians and democratisers who bring science to the people as a downward (and thereby elevating) movement. To draw on the vernacular of the publishing business, the *cross-over genres* of scientific representations, those in which the scientific expert constructs articulations for a lay readership, provide a particularly edifying and rich illustration of the narrative structuring of the scientific imagination. Here, even as scientific practice (and practitioners) must necessarily provide the source-point for cultural inscription, wider commonsenses must also be tapped to provide an intelligible reference point for the bridging (and delineation) of worlds.

The literature I have selected for close study in this chapter belongs to a distinctive strand within popular writing about science. These are typified by their designated readership (highly educated, non-, or not necessarily, scientific or medical but perhaps, otherwise, professional); the credentials of their authors (often well-celebrated scientists); their often glossy and very tasteful

---

8 Michael Crichton and Robin Cook (both medical doctors by training) are prominent examples of that authorial crossover, although their fictional works could not be described as exactly celebratory of science. Crichton's *Jurassic Park* and *The Lost World*, for example, provide a significant critique of the epistemic economies of recombinant genetic science, yet this critique is articulated through an alternative scientific perspective (chaos theory) embodied in the character of Ian Malcolm.

production values (with pricing in the upper range, typical of academic books); and their availability in the 'better' bookstores, usually in the 'general science' and 'general medicine' sections (separate, but not far from the textbooks) and, not infrequently (depending on the celebrity of the author), as special displays at the front of the store. The overwhelming majority of this literature takes a laudatory approach to scientific progress.[9] Here, critique, where it appears, tends to be weighted against projections and estimations of the overwhelming benefits of scientific progress. Rationales for new scientific directions are characteristically expressed in terms of noble goals – the relief of suffering; the quest for knowledge; the forging, in deferential acknowledgement, of the human capacity to grasp and to harness the (recalcitrant/magnificent) forces of nature.

This chapter focuses on three early *cross-over* genetics advocacy texts which occupy the wider generic literature described above, and that both characterised and indeed set the terms for a proliferating field of progress-orientated texts on the topic of genes and genetics. All three of the selected texts make a case for a widening practice of recombinant genetics. All, interestingly, situate their genetics rationales within a broader, albeit many versioned, liberal-egalitarian political vision. All also cull distinctive cultural narratives, metaphors and commonsenses that in many respects hearken back to the earliest period of modernist thinking, to articulate utopian projections of the genetics revolution. Stephanie Yanchinski's *Setting Genes to Work* (1985), the earliest text, presents a libertarian, industrial utopia. Capitalist metaphors and narratives dominate the text in an imagined world of perpetual production (of food, chemicals and wealth), a molecular matrix in which discourses of class are both displaced and reinvested in the micro-bodies – the new labouring masses – of the cell. In *Genetics for Beginners* (1993) Steve Jones, with illustrator Borin Van Loon, present what might be termed a liberal imperialist utopia. Here we find a textual and visualised colonial adventure narrative in which the biological landscape is mapped, its resources plundered and the scientist becomes hunter/civiliser of rogue genetic nature(s). Finally, there is *The Lives to Come: The Genetic Revolution and Human Possibilities* (1996), Philip Kitcher's diagnostic utopia in which futures, corporeally mapped, are located in the boundaried, yet metamorphic bodies of the genes. Here is presented the seductive projection of a rational, democratic pharmocracy in which freedom is to be found in the disciplinary regimes of medical diagnosis and rationally planned – 'utopian eugenic', in Kitcher's terminology – reproductions.

---

9 There is, however, a marginal sub-genre of such texts which provide critiques of science and medicine (e.g. Rose et al. 1984; Elkington 1985; Piller and Yamamoto 1988; Garrett 1994; Rose and Rose 2012). These texts appear infrequently, rarely in numbers greater than two or three at any one time (quite often there are none at all).

Monads and Machines: Yanchinski's Industrial Utopia

monad: (n.) the number one; a unit; an ultimate unit of being, material and psychical; spirit; God; a hypothetical primitive living organism or unit of organic life.[10]

In *Volatile Bodies* (1994), Elizabeth Grosz examines Spinoza's concept of the monad as a radical challenge to the disunification of body and agency characteristic of the Western Cartesian tradition. The monad constitutes singularity of process, matter and purpose; each attribute or entity only part of the infinitely possible expressions of nature. Grosz argues that monism thus 'frees the body' from two key aspects of Cartesian thought: its dualism and its 'dominant mechanistic models and metaphors' (p. 11). In Spinoza's conceptualisation of a cell, for example, there is no pre-given structure, animated by an external energy source (as fuel added to a machine), but rather its very being-ness *as* a cell arises from a self-generating, continuous metabolic process:

> The forms of determinatedness, temporal and historical continuity, and the relations a thing has with coexisting things provide the entity with its identity. Its unity is not a function of its machine operations as a closed system (i.e. its functional integration) but arises from a sustained sequence of states in a unified plurality. (pp. 12–13)

Like the atom before it, the commonsense constitution of the gene as a self circumscribed and singular unit of existence bespeaks a monistic unification of matter and energy, meaning and substance. Yet paradoxically, the manipulations of recombinant genetics would also seem to *reinvest* the putative monadic bodies of genes back into the dualistic traditions of modern scientific thought and practice. With recombinant genetics the body of the gene is both conscripted for, and understood to be (re)animated through, the intentionalities of the scientist-engineer and the invented body-mechanics of the recombinant organism. This contradictory fusing of monad and machine is tellingly articulated through the linguistic, organic and industrial metaphors that permeate the dominant literatures, scientific[11] and popular, of contemporary genetics and that are explicitly elaborated in Yanchinski's utopian assessment of recombinant genetics as heralding a new industrial era.

---

10 *Chambers English Dictionary*. 1990. Edinburgh: Chambers.

11 Elsewhere (Steinberg 1997), I have discussed the constitution of the gene as monad in machine within both professional and popular literatures on preimplantation diagnosis (genetic biopsy on embryos produced through *in-vitro* fertilisation).

Worker Bodies and the Industrial Metaphor

> We are in the middle of another industrial revolution, and most people are only just realizing it. This industrial revolution, called "biotechnology" depends not on iron and steel but mainly on microbes which scientists are converting into minuscule factories for exotic drugs, industrial chemicals, fuel and even food. The "bio" in biotechnology refers to bacteria and yeast, mainly, but also to other living cells such as plants, fungi and algae. The "technology" consists of batteries of *gleaming steel vats* full of microbes … [and] hundreds of valves opening and closing to computer-set rhythms, guided by artificial intelligence – all the trappings of our electronics world applied to keeping these microbes producing at *top efficiency*. (Yanchinski 1985, p. 7, my emphasis)

Encapsulated in the notion of 'setting genes to work', the title of Yanchinski's appraisal of recombinant genetics, are the key connotations that frame not only her own projected narrative of a new industrial utopia, but also, as we shall see, inform the narrative economies of the later texts by Jones and Van Loon (1993) and Kitcher (1996). First is the emphasis on work. In Yanchinski's formulation, the significance of genes lies in their perceived productive capacities and, perhaps more to the point, in their projective constitution as *labouring entities*. The notion of 'setting' is also suggestive of a reading that genes do not work *unless* 'set'. Here the gene is, at least in part, connotatively evoked as an idle, undisciplined body, a potential yet resistant energy, animated into its proper function only by an outside agency (the 'setter'). Such a construction is suggestively reminiscent of Victorian constructions of a recalcitrant working class, defined by the contrast of its muscle to its mind to industry, efficiency and economy. Congruent with industrial logics located in more conventional settings, genes as workers are implicitly constituted as docile[12] and yet resistant, as performative objects and objectified subjects in a disciplinary economy that they embody but do not own. Central, then, to Yanchinski's utopian prognostication of mastered microbes in 'gleaming steel' vat-factories are genes as imagined bodies, nostalgically and narratively constituted through a discourse of class and a commercial rationality.

The Gene as Class Ideal

> So yeasts and bacteria are the *new labouring masses* toiling to make the wheels of the biotechnology industry run. (Yanchinski 1985, p. 20, my emphasis)

---

12 Foucault (1977). See also Steinberg (1996b; 1997) for further discussion of genes as 'docile' bodies in the Foucauldian sense.

A pervasive linchpin of Yanchinski's industrial narrative is articulated through the metaphors of industrial capitalism. Indeed, the language of class permeates Yanchinski's idealisation of the gene in both its projected commercial and medical applications.[13] Languages of 'toiling labouring masses', as above; of 'aristocratic bosses'[14] (ibid.); 'factory foremen' (p. 27);[15] and 'gene machines' (p. 49)[16] essentialise genes as mechanised components of the transgenic factory.[17] In these projections, class divisions evocative of Blake's 'dark Satanic mills' are reconstituted – recuperated and transplanted but still, apparently, intact. The projection of class referents and identities onto the gene has a double movement – naturalising class divisions and yet displacing them onto putatively lesser natures. Implicit in Yanchinski's genetified factory is a promise of human freedom from class oppression even as such inequalities are recuperated for human investment, conceptual and commercial.

---

13 Yanchinski's discussion of using genetically engineered *E. coli* to produce Human Growth Hormone provides an edifying illustration of the inter-embeddedness of commercial and medicalised logics (and of the genetic diagnostic imperialism discussed in the previous chapter) in her bio-industrial utopia: 'Human growth hormone is constantly in demand for treating dwarfism, and currently most of it comes from the brains of human cadavers. Unlimited supplies of this scarce material would not only ensure that all dwarfs received treatment, but would also enable doctors to consider treating 'short stature' children who are simply undersized for their age and help heal bone fractures. New material, new niches' (p. 102).

14 As Yanchinski puts it: '[Genetically manipulated] bacteria and yeasts are highly specialised and overbred, the aristocrats of the microbe world, and they thrive only in the highly artificial environment of the fermenter. Change the temperature by a degree or two, or the acidity by a fraction, or delay adding a food component, and they react badly, often reverting back to the wild type or simply dying' (p. 20). In addition to the class discourse through which it is science that confers 'aristocratic' identity to the gene, there is also colonial subtext embedded in the evolutionary notion of 'wild types', those primitive entities to which the cultured gene reverts without the civilising accoutrements of the laboratory hot house.

15 Yanchinski reiterates an early conceptualisation, in the history of genetic science, that 'RNA was the "factory foreman" who directed the manufacture of proteins by transcribing the *blueprint* laid down by the "boss" DNA' (p. 27).

16 Yanchinski writes: 'Clever chemical engineers … incorporated the bench-top chemistry into a range of "gene machines" which busy scientists not so skilled in chemistry could buy to make gene fragments' (p. 49).

17 'The fourth biotechnology era can be said to have begun with the advent of genetic engineering in the early 1970s. The tremendous impact of this technology, which involves inserting foreign genes into microbial cells, thus converting the host into a protein-making factory, will only be truly felt in twenty years' time' (Yanchinski 1985, p. 15).

Similarly, the construction of labouring agency is contradictory, on the one hand located in the *self-organising* capacities and monadic properties attributed to genes, and yet at the same time, utilised as functionary cogs in the wheel of a superordinate disciplinary economy. Factory discipline itself is invoked both through the language of intimate precision (to map, cut and paste) attributed to the capabilities of the genetic engineer, and through the idealisation, accruing to such language, of productive efficiency. Weiss (1987) has argued that a preoccupation with social productivity and rational selection articulates a technocratic-managerial logic that was a hallmark of early eugenic philosophy.[18] In Yanchinski's utopia, as with the languages of class, the displacement of the language of selective breeding on to genes recuperates (even as it ostensibly promises distance or liberation from) its historical resonances of productive and reproductive 'fitness', national efficiency and racial hygiene.[19]

To summarise then, the ambivalent construction of a self-governing yet externally governed, self-assembled yet scientifically recombined, production processes and producers is evocative of industrial logics and narratives hearkening back to the nineteenth century. Indeed, Yanchinski's idealisation of 'immobilised cell technology' (p. 24)[20] epitomises precisely the industrial fantasy of continuous production and perpetually renewable (and essentially expendable) labouring forces.

Dark Continents: Genetics as Colonial Quest

Heroic narratives have been given a particular inflection in discourses of the nation generated since the emergence of the nation-state in early-modern Europe. (Dawson 1994, p. 1)

Even as Yanchinski's industrial utopia, in its recuperation of the monad for the machine, focuses on the *gene*, the narrative sweep of Jones' and Van Loon's *Genetics for Beginners* is focused on the genetic *scientist*. Here, as we shall see, the

---

18 Weiss (1987) examines efficiency as a class ideal and industrial logic underpinning eugenic politics as articulated through the philosophy of William Schallmayer.

19 Weiss further argues that the logic of efficiency 'formed the common bond which united those race hygienists who accepted ideologies of Aryan supremacy and those non-racist eugenicists, like Schallmayer, who vehemently rejected them' (p. 6).

20 Of 'immobilized cell technology', Yanchinski writes: 'But most exciting of all is the thought of … cells, instead of floating suspended in their own food [but rather] fixed to a solid support – in a tall steel column say, through which flowed a fine stream bearing all the food nutrients the cells need. If the column were cheap enough, it could even be thrown away when the cells were past further use' (p. 24).

geneticist as heroic (if modest) adventurer in the uncharted terrains of the molecular world is textually embedded in a repertoire of eugenic and colonial imagery that is, again, culled from the Victorian period. Of particular salience in this context are the distinctive motifs – surrounding nation, race and sexuality – which can be traced through organising narratives of genetic cartography and conquest.[21]

Colonial Cartographies: Mapping the Gene as Foreign Landscape

> In spite of the new confusion in genetics, it is clear that – just as the first explorations of South America – hidden within the genetic map there are new and startling facts about genes, about disease and about evolution. Now there is a scheme for the great Map of Ourselves – a list of the three thousand million letters in the human DNA. It may be complete by the year 2000.
>
> It will cost a lot: but like many maps will be the first step to exploiting the country which it reveals. (Jones and Van Loon 1993, p. 112)

The textual investment of genetics in analogies of colonial exploitation is not incidental to Jones and Van Loon's imagining of the biotechnology revolution. *Genetics for Beginners* is peppered with images of white colonial hunters, treacherous African and jungle landscapes, boys-own stories of treasure hunts[22] and Lone Rangers supported by backward, but nevertheless complicit, natives. Incorporated in such images are constructions of both the geneticist and the enterprise of genetics within the recognisable conventions of the Victorian quest narrative,[23] with their themes of dark continents to be taken, known and ruled, masculine aspiration and racialised–sexualised imperial fantasy (see Figure 2.1).

---

21 Steve Jones, as I discuss further in Chapter 3, is a prominent scientist cum public figure in Britain and one of the UK's most important popularisers of genetics. *Genetics For Beginners* is the most populist of his multiple works on the social, scientific and cultural importance of the gene and genetics, which span a spectrum of populist to high-brow. The *For Beginners* convention (used by a number of publishers) used comic strip renderings to explain academic fields, theories or movements: *Philosophy for Beginners* (Osbourn and Edne 2005); *Marx for Beginners* (Rius 2003); *Postmodernism for Beginners* (Powell 1997). The humour of the series lies in part in the intended visual misfit of populist form with high-brow subject matter (with the high-brow as the object of the joke). The series has also been arguably intended as an educational tool to make accessible theories and fields that are normally the preserve of experts.

22 Jones writes: '[S]ome biologists believe that it is worth looking for treasure in the depths of the molecular forest – after all, no-one has any idea what could be hidden there' (p. 125).

23 See, for example, Showalter (1992); Dawson (1994).

Figure 2.1

Source: Jones, S. and B. Van Loon. 2011. *Introducing Genetics*. London: Icon Books

The notion of gene-space as a landscape, for example, incorporates Yanchinski's monadic-machine bodies in a subtextual motif of rational domestication. The putative African or jungle genes, for example, are visualised within the conventions of dominant racialised mythologies of the rogue primitive, whose very nativeness seems to demand the cartographic and conquering interventions of white civilisation. The genetic landscape is valued for the dormant, hidden riches of which its native inhabitants seem constituted as both unaware and/or intrinsically incapable of exploiting. The quest for the 'dark continent' is, furthermore, organised around a particular imperial-commercial logic; it is the presumptive institution of Western know-how that locates, cultivates and expropriates the latent economic potential understood to be embedded in the 'alien' landscape.

Civilising Missions: Metaphoric Economies of Nation, Race and Sexuality

If the 'foreign' imaginary discussed above is often coded through controlling myths[24] of colonial Africa and South America, so too is the 'familiar', the imagined genetic coloniser, predominantly visualised through dominant mythologies of white, male, upper class Britishness.

The 'I say, Carruthers' of the generic geneticist of Figure 2.2, for example, melds with the intended humour in Figure 2.3 that re-presents Watson and Crick as Holmes and Watson (p. 57). Both representations constitute visual puns grounded in a distinctive version of white British masculinity articulated through its purported 'quaintness', its bumbling (yet incisive) modesty, its idiosyncratic (even ironic) genius and, perhaps most importantly, its taken-for-granted privilege of both entry and ownership.

An investment of genetics in notions of the British nation (and white British masculinity) is also suggestively forged through other representative clichés. Taking up the common theme discussed above of 'nonsense DNA', Jones and Van Loon take up the Chinese language as a metaphor of generic-genetic incomprehensibility:

> It was as if the owner's manual for an English car was interrupted by sentences in Chinese which had to be nipped out before the instructions could be read properly. (p. 90)

---

24 Patricia Hill Collins (1990) uses the term 'controlling myths' to describe the dominant cultural mythologies invested in commonsense constructions of African American maleness and femaleness. In particular, she locates these mythologies within the specific history of European colonial exploitation of the African continent and the enslavement of African peoples.

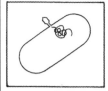 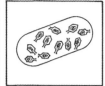 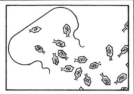

But how could such a simple substance copy itself and pass information from one generation to the next?

There was already a hint. There were only four units in DNA — Adenine, Guanine, Cytosine and Thymine (A, G, C and T for short). The number varied from creature to creature — but there was always the same proportion of A to T and G to C.

53

Figure 2.2

Source: Jones, S. and B. Van Loon. 2011. *Introducing Genetics.* London: Icon Books

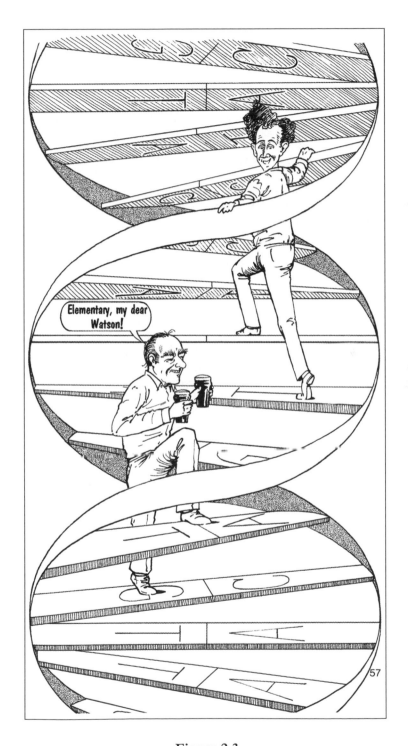

Figure 2.3

Source: Jones, S. and B. Van Loon. 2011. *Introducing Genetics.* London: Icon Books

But then — the first of the big surprises. Suddenly, a lot of the structure of the haemoglobin and other genes began to make no sense at all. Mendel would have hated it!

Reading along the DNA message showed that there was far more DNA within each gene than was needed to make the protein.

The way the gene worked was even stranger. The whole of the DNA of, for example, the ß-chain was read off into a long messenger RNA, but then — astonishingly — sections of the message were cut out and thrown away. An edited version of the message left the nucleus.

It was as if the owner's manual for an English car was interrupted by sentences in Chinese which had to be snipped out before the instructions could be read properly.

Chock the rear wheels, jack up the front of the car and support on の体格を忠実に再現させることになる。axle stands located between the longitudinal members. Move the selector lever to the それはあたかも、生物体が一連の議会選挙区に分けられて、支持する者を当選させるために、各選挙区から代表団が送られてくるかのようだ。この 'D' position. Undo and remove the bolt securing the transmission control cable retainer to the casing. Undo the two control cable adjustment locking nuts and pull ジェミュールは血流に送りこまれる。それから生殖細胞に再び集まり、親 ダーウィンにとってこの仮説は、獲得形質の遺伝をみごとに説明するものだった。たとえばある生物が自分の努力で手足の筋肉を大きくすると、そthe outer の発達した部分からcables出るジェミュールfrom the transmission casing. The の数がふえ、生殖細胞に多く集まる。control inner cable may したがって、now be disconnected from the valve block detent rod and the park その生物の努力の実りが、自動的に子孫に遺伝するのだ。lockrod. 逆に、もし手や足などの器官を使わないでいると、そこから出るジェミュ Make a ールの数は減少し、その不足もまた子に遺伝する。この説は、進化における偶発的変異の役割を予期していた人々に強い確信をもたせた。そして note ダーウィン自身がこのような説を容認したという事実は、ラマルク説を再 評価させる要因になった。1870年から、ラマルクの説の再検討が本格化し of the electrical cable connections at the starter た。そしてダーウィンが死ぬ少し前の数年のうちに、ものすごい勢いのラinhibitor switch and detach the cables. The front of the car should now be lowered to the ground. The weight of the car should now be ──── unit must now be taken マルク説復活が起こった。

90

Figure 2.4

Source: Jones, S. and B. Van Loon. 2011. *Introducing Genetics.* London: Icon Books

Englishness thus constitutes not only the presumptive reader and owner of the sought code (and the functional apparatus to which it refers), but the generic properties of functional productivity – Englishness, in other words, is deployed here as a metaphor for meaningfulness itself. Chinese, by contrast, is constituted as intrinsically unreadable, as alien interloper, needing to be 'nipped out' for proper order to be restored.

A similar Orientalism is also evoked in Jones and Van Loon's representation of boy-preference (constituted as a potential problem accruing from the use of genetic techniques for sex selection) as a function of the putative underdevelopment of Indian culture (see Figure 2.5).

Here, the ownership of genetics is graphically visualised as white, male and Western. India is constituted through the recognisable Orientalist mythology of hyper-patriarchy; its 'extreme' oppression of women iconographic of its 'backwardness'[25] and counterposed to the more enlightened anti-sexism implicitly attributed to British culture.

The racialised imagery of nation that characterises Jones and Van Loon's colonial genetics adventure is mediated through explicitly eugenic textual economies.

Figures 2.1 and 2.6, for example draw explicitly on what might be described as generic eugenic imagery. There is a double movement here as, for example, Figure 2.1's image of the 'white man wrecked on an island of Negros' is taken up in what might be read as an intended critique of the simplistic wrong-headedness of Victorian eugenics/genetics: '[a] highly favoured white cannot blanch a nation of Negros'. Yet the heternormative assumption of male access – 'civilised' as opposed to 'savage' – that is characteristic of colonial conquest narratives is reasserted rather than rejected in the white British man's putatively 'humorous' directive 'Ok you chaps let me have the pick of your wives!! I *am* British after all … '. Masculinity is the subject, not the object of the joke here. Similarly, what might be described as the classic imagery of constitutional 'degeneracy' evoked in Figure 2.6 is (re)deployed, again 'humorously', possibly intended as an 'ironic commentary', on the ways in which mechanisms of inheritance can be recognised in the face of one's offspring. Yet the plundering of racist imagery and language to challenge racism is disingenuous at best. The ironic inflection seemingly intended in the use of this imagery does not work precisely because the images deployed were products of a eugenic colonial enterprise, in which genetics was an

---

25 See, for example, Strobel (1991) for a discussion of the ways in which imperial conquest in India was, in part, rationalised through references to Indian women's status as index of its putatively lesser civilisation.

Figure 2.5

Source: Jones, S. and B. Van Loon. 2011. *Introducing Genetics*. London: Icon Books

Source: Jones, S. and B. Van Loon. 2011. *Introducing Genetics*. London: Icon Books

Figure 2.6

organising discourse.[26] However ironic their motivations, Jones and Van Loon's choice to take up such imagery as a recognisable reference point for the lay-reader bespeaks the continued currency of eugenic-colonial narratives in both popular and scientific commonsense. In turn, contemporary genetics is revealed as a hospitable, indeed generative, context for the recuperation of such narratives.

Making the World Safe for Genetics: Kitcher's Diagnostic Democracy

> Geneticists warmed to their priestly role with ease. The new industrial order had elevated practitioners of the physical sciences to positions of power and public service. (Kevles 1985, p. 69)

If Victorian discourses of industrial capitalism and colonial adventure characterise the early utopian imaginings of recombinant genetics in the previous texts, their diagnostic implications are elaborately realised in Paul Kitcher's *The Lives to Come*. Here the geneticist as heroic explorer, divining the mysteries of molecular space, also serves a reminiscently 'priestly role' in a narratively projected recuperation of both the physical and social world. In Kitcher's imaginary, genetics is constituted as both a journey, inflected with biblical import, and a grail of modernist rationality applied both to reproductive practice – what Kitcher terms 'utopian eugenics' – and to the forging of a new social order of diagnostic discipline.

Of Angels Bearing Gifts

> Another image of a great journey recurs in advertisements for contemporary biomedical research. No less a figure than Walter Gilbert, Nobel Laureate and co-inventor of sequencing technology, has seen genome sequencing not as tedious drudgery, but knightly enterprise. The sequence of the human genome is the "Grail". (Kitcher 1996, p. 88)

---

26 A similarly disturbing and suggestively anti-Semitic reference is taken up in the visual 'gag' deployed to critique Lamark's genetic experiments (cutting off the tails of mice to test his hypothesis that genetic traits are acquired environmentally). The figure of Lamark, tossing amputated mouse tails into a 'tail pail', laments: 'This is the twentieth generation and – damn it – still they've all got tails!'. The wifely figure of a woman in the background replies: 'But Jews have been doing the same thing for years'. The textual juxtaposition of Jews and tails, and the veiled reference to ritual circumcision recuperates racialised mythologies of difference/deviance embodied, in part, through mutilation. See, for example, Gilman (1991) for further discussion of the Jew as deviance embodied, and the role of circumcision in that corporealisation.

They journey in hope, not knowing for sure that they are playing Christian to the naysayers' Pliable, whether they are Arthurian knights or Don Quixote. (p. 89)

If there were a sequence-bearing angel, what use could we make of the gift? (Ibid., p. 91)

The allegory of the masculine quest, divinely inspired, ennobling and edifying provides the dominant narrative frame for Kitcher's construction of the social importance of both the geneticist and his science. An entire chapter is given over to a sequence of road allegories characterised not only by their quasi-religious overtones but by their investment in a particular version of masculinity: chivalric, exalted, transcendent and, significantly, seductive. The reader is given, for example, John Bunyan's pilgrim Christian who 'struggles through the mire, goes on to face many other reversals, and ultimately wins his heavenly reward' (p. 21). Christian is the visionary genetic journeyman, his eyes beyond the earthly limitations and 'short term difficulties' (ibid.) of genetic testing, to the 'eventual hope … to be able to intervene, to treat to cure' (ibid.). As an edifying tale, Christian's journey is taken up as a defensive fable. In his counterpoint to the 'naysayers' Pliable, who loses faith at the first impasse – that genetic testing 'can predict when disease will strike … [but is] … helpless to avert or alleviate it' (ibid.) – Christian, as 'visionary', embodies the moral high ground (and moral of the story) of genetics' noble purpose, in which the wonder is focused not on mundane pragmatic outcomes, but in the great journey itself. Similarly, narratives of knightly quest and angelic prophets bearing gifts, 'maps of destiny' (p. 90) and 'royal roads' (p. 91) imbue a romanticised, metanymic divinity to each localised technical mastery that signposts the molecular path to enlightenment.

The rhetoric of noble purpose – to alleviate suffering – common to all these narratives moreover, implicitly articulates a rescue fantasy that, as Judith Williamson (1997) has suggested, not only 'runs back to medieval chivalry', but in which eroticised projections of 'men doing valiant deeds as a way of courting and "serving" women' are encoded. As romantic saviour, the persona of the geneticist is not only prefigured male, but is embedded in a strictly heteronormative narrative frame. For example, the evocation of the knightly conqueror, indeed even of the romantic outlaw Don Quixote, presents a reminiscent, though perhaps less obviously racialised, heterosexual subtext to the narratives (and penetrative symbolic economies) of colonial conquest taken up by Jones and Van Loon. In this context, genetic science is complexly engendered. On the one hand, genetics is graphically masculinised, both as a male preserve and as a journey to a masculinity exalted just short of godhead. At the same time, genetic science is narratively positioned as endangered by, for example, the 'naysayers' pessimism – an implicitly feminised construction within the dominant logic of the genre. Indeed, it can be argued that the construction of genetics as 'grail' evokes comparable symbologies of masculine virtue – liberty and justice, for example – that are characteristically embodied female. In turn, the relationship of masculine rescuer to endangered science offers preferred positions of identification in similarly gendered terms. For it is clearly a masculine agency and eros that is imagined as prime mover of both the science and its narratives; dangerous journeys both, to be braved by brave men. The framing of genetics through romantic fantasies of masculine endeavour clearly link the gendered, heterosexualised imaginaries of Kitcher and Jones and Van Loon. At the same time, there are revealing distinctions to be drawn between narratives of quest and of conquest. If the latter evokes a manful assertion (and mans-own world) of rugged derring-do, the former posits the recuperated conqueror, now protector and humane husband of fragile, but fantastical, futures.

Diagnostic Designs

If the romantic quest narrative signals the cosmological import, as it were, of the genetics journey envisaged by Kitcher, more pragmatic imaginings frame the forging of a newly genetified social order. Two themes in particular emerge as significant in this context: the primary location of genetified futures and rationales in a medicalised logic and the articulation of its recuperative potentials in a project of 'safe' eugenics. It is in this context that significant distinctions can be drawn between this and the previous two accounts of the genetics revolution. While all three texts, for example, do note problems with genetics (for example accruing to its historical resonances with 'stigmatised' eugenics), Kitcher's concern for the potential for harm emergent in contemporary molecular biology is not a subsidiary focus of the text. Indeed, the alleviation of suffering and the cultivation of an egalitarian society are higher purposes he pointedly invests in the potentials of a genetified future and are themes sustained through (and perhaps behind) the allegorical narratives of noble journeys and protective knights. Secondly, although diagnostic implications for the new technologies are taken up by Yanchinski and Jones and Van Loon, this is chiefly as an adjunct to, and incorporated within, their overarching and respective industrial and conquest logics. By contrast, as has been discussed earlier, while Kitcher's text takes up key metaphoric and narrational economies of the previous texts, it is the diagnostic-medical applications of genetics that are characterised as the dangerous and yet ultimately redeeming centre of the road to molecular enlightenment.

Testing Times

> Just after delivery, the nurses took a sample of the baby's blood, and in accordance with the medical standards of 2020, the pediatrician will now give the parents the results of the analysis, the "genetic report card". The interview proves more

frightening in prospect than in actuality. Although the long columns of statistics are initially baffling, the doctor points out that most of the risks are normal, or below normal, and that the only worrying figures are somewhat elevated probabilities of diabetes and hypertension. She recommends attention to diet from an early age, as well as regular checks of blood sugar levels. Like many others who have experienced the new medicine, the parents are grateful they can take rational steps to promote their child's health. (Kitcher 1996, p. 23)

The 'genetics report card' story both leads off and encapsulates Kitcher's utopian imaginings of a world where genetics is the centre-point around which social life is to be rationally re-ordered. As this excerpt suggests, there are a number of key assumptions underpinning the diagnostic logic of a world defined by genetic testing. First is the unilateral investment in scientific expertise, whose assumptions can be summarised in a neat narrative arc: genetic science produces accurate facts; doctors explain these facts to patients, who are in turn grateful; and 'rational steps' can then be (and are) taken to accommodate the genetic propensities revealed. Second is the movement towards universalisation of testing. Indeed, as I discuss further below, Kitcher explores the problematic, but ultimately, in his view, justified projection of genetic screening into progressively widening spheres of social life, from the pre- to post-conceptive, and from the workplace to health and life insurance, to the criminal justice system. In this context, the privileging of medical scientific discourse and agency reaches its apotheosis as genetic diagnostic logics are taken to impel a radical reconstitution of both social order and cultural commonsense.

Thirdly, the notion of 'rational steps' is itself invested in the presumption of (expertly read) singularities characteristic of gene discourse more generally. In the scenario outlined above, for example, it is imagined that a singular 'problematic' predisposition will be the outcome of a battery of genetic testing for which a set of appropriate responses will be clearly available. While Kitcher acknowledges the problem of assuming that all testing will involve 'all of the features that make the perfect test so attractive' (p. 24), his critique never considers either the probability that widening ranges of genetic testing will produce multiple projected propensities, or that the 'rational steps' taken (if indeed there are any at all) for one condition or predisposition may conflict with those to be taken for another. Similarly, while Kitcher notes the problem of 'a world of tests without therapies', his solution is a redoubled investment in genetics: 'the best remedy may be to double our research efforts' (ibid.).

Obscured in Kitcher's idealisation of the possibility that 'we shall be able to test at will – fetuses, couples who plan to marry, people who are sick, and those who show no symptoms' (p. 25) is a another story: a dystopia in which 'our' captivation with genetics translates into captivity as we become subject to a relentless and deeply contradictory disciplinary regime of genetic diagnosis.

Read this way, Kitcher's utopia evokes not a rational, healthful and egalitarian society, but rather an encroaching, invasive medicalisation of life – what Illich (1976) termed *diagnostic imperialism* and Lemke (2004) termed *genetic governmentality* – in which clinical, social and cultural iatrogenic effects (clinical harm, social dependency on experts, and the loss of belief in ones ability to cope with illness or pain) inevitably accrue.

Eugenic Recuperations

> Once we have left the garden of genetic innocence, some form of eugenics is inescapable, and our first task must be to discover where among the available options we can find the safest home. (p. 204)

> As a theoretical discipline, eugenics responds to our convictions that it is irresponsible not to do what can be done to prevent deep human suffering, yet it must face the challenge of showing that its claims about the values of lives are not the arrogant judgements of an elite group. (Kitcher 1996, p. 192)

Kitcher's elaboration of the problems that may accrue to the expansion of genetic testing revolve around its perceived potential to exacerbate existing inequalities – i.e. 'unsafe' eugenics. In this context, he projects a number of dystopian scenarios: including workplace to health and life insurance discrimination; the intensification of class and racialised divisions (which can accrue from diagnostic errors (p. 33) or coercion); a world, as has been mentioned, of tests without therapies and with insufficient demographic and etiological statistics to make sense of diagnostic data; and the problematic assumption of rationality and rational choice as the prime movers of science and of clinician and patient agency. The answer to these rather overwhelming problems, Kitcher suggests, are to be found in the delicate balancing act of 'utopian eugenics':

> Utopian eugenics would use reliable genetic information in prenatal tests that would be available equally to all citizens. Although there would be widespread public discussion of values and of the social consequences of individual decisions, there would be no societally imposed restrictions on reproductive choices – citizens would be educated but not coerced. Finally, there would be universally shared respect for difference coupled with a public commitment to realizing the potential of all those who are born. (p. 202)

What is perhaps most striking about 'utopian eugenics' is the way in which it is predicated on the clearly sincere but nevertheless facile projective assertion of some future equality. We are to assume that genetic screening will not serve only the privileged even as it presages a new class order dividing a small elite of experts from 'grateful' lay-patients. We are to assume that it will not be coercive while at the same time, forced testing in the workplace and in reproductive contexts, according to Kitcher's own assessment, should (would) be regarded as justified where 'risks' to third parties are involved. We are to assume that 'accurate and adequate information' (about epidemiological probabilities, and the full range of 'environmental factors relevant to the expression of disease') – what Kitcher characterises as 'mountains of statistics' (p. 72) – can simply, albeit with 'considerable ingenuity and immense labour', be provided in order that rational decisions can be made. We are to assume that adequate and independent counselling, 'responsive to the [social and individual] predicaments of all citizens' (p. 86) and fully cognisant of that 'mountain of statistics', can be devised. We are to assume that existing stigmas associated with illness (and social inequality) will be eradicated and that differences constituted through 'genetic disclosures will "spark sympathy" rather than lead to social discrimination' (p. 151). Indeed we are to assume that 'systematic knowledge of the human genome could correct prejudices ... that the catalogue of genetic disabilities is distributed across racial and ethnic groups in roughly equal fashion, engendering the conviction that while there are differences, taken as a whole, all groups are 'genetically equal' (p. 154).

In this context a significant contradiction emerges. On the one hand, a drift to inequality is seen as inherent in genetic screening. At the same time, genetic science is held to herald substantive potentials for democratisation. The extraordinary number of sweeping 'ifs' that must be satisfied in order that genetics does *not* reinforce inequalities, indeed, in order to make a world safe for genetics, is a testament to the tragic absurdity of such wishful thinking. Kitcher argues that 'without attention to the social surroundings in which molecular medicine is practiced [developments in genetic technology] would simply magnify inequalities that are already present' (p. 97). Yet, at the same time, it is in genetics' very propensity to fuel oppressions that Kitcher, rather disturbingly, perceives a bedrock (impetus?) from which such social changes will emerge. Genetics, in this formulation, is both embedded in and yet disaggregated from social relations. Thus we are to understand that while genetic screening will intensify oppressions, it will also transcend them. Perhaps more to the point, the investment of such faith in the utopian fantasy of 2020 seems premised on yet another singularity. This is the notion that a science produced *out* of, and seen as desirable *within*, an unequal social order, would continue to make sense, and for the same reasons, in a profoundly altered social context. Indeed, there seems a further assumption that such a science's acknowledged propensities to do harm would not *undo* a democracy. Kitcher states: '[f]rom the beginning, it will be crucial to prepare the right social settings for the new applications of molecular medicine' (p. 182). Yet as even at the time of Kitcher's writing, we were well past the beginnings of this project, a rhetoric of good intentions can only be both disquieting and duplicitous.

In the Imaginarium of Genes

In its examination of the narrative framing of these selected texts from a wider genre of popular-scientific advocacy of recombinant genetics, this chapter has closely examined the inter-embeddedness of scientific and popular commonsense. The issue raised is not, as Sontag's early work suggested, that science is corrupted by metaphor, but rather that the *particular* metaphors invoked *situate* particular scientific perspectives and practices in ways demanding critical evaluation. Perhaps what is most significant in the context of this study, is the way in which these narratives reveal the imagined bodies and communities constituted through genetic discourse as both grounded in and contingent upon conditions of social inequality. Common to all of these utopian narratives are investments in elite (classed, gendered and racialised) knowledges and discursive practices that have been historically implicated in both the (re)production and normalisation of social divisions. Indeed, all invest the agency of both social stability and social change in scientific 'readers' and 'writers' – in both the experimental and representational senses of the terms. Similarly, all of these narratives articulate a slight-of-hand through which genetics is, on the one hand, identified with institutionally oppressive structures (dark satanic mills, colonial invasion, diagnostic imperialism) and yet, on the other, projectively reconstructed as rescuer. The liberal-minded recuperation, for example, of the 'bad' eugenic associations attendant upon recombinant genetics is narratively redeemed through assertive ascriptions of marvellous potentials and noble intent. What is revealed in this context is the extraordinary power of the redemption narrative, whether framed as adventure or romance or pragmatic projection. For it is this narrational frame that makes it possible for the overwhelming dystopian dimensions of the genetics revolution to be acknowledged, embattled and ultimately, saved.

Chapter 3
Genes and the (Post)Racial Imaginary[1]

Introduction

Since its inception in the 1970s, developments in recombinant DNA technology have progressively transformed the constitutive practices of whole fields from medicine to industry to agriculture to criminalistics, with contiguous transformations of our languages and commonsense understandings of disease, kinship and identity. In turn, the gene has become a dominant cultural referent not only for processes of biological, but also of *social* reproduction, and a key cultural metaphor for a range of social relations, including, perhaps most ambivalently, for 'race', nation and otherwise imagined bodies and communities.[2] Moreover, the entry of the language of genes into popular discourse has crossed representational genres from documentary reportage to science fiction, from textbook to comic strip, from metaphor to gag.[3] An emergent strand of critical writing about the inter-relationships of professional and popular spheres of scientific commonsense have emphasised both the narrative character of scientific discourse and of the role of the scientist as author in both contexts. Indeed, as I suggest in the previous chapter, an examination of moments of popularisation with respect to particular scientific enterprises can reveal the ways in which the social relations and conceptual trajectories of scientific cultures shape and are shaped by broader popular discourses.

---

1 An earlier version of this chapter appeared in Avtar Brah and Annie Coombs (2000) *Hybridity and its Discontents: Politics, Science, Culture*. London: Routledge.

2 Anderson (1982). See also Hall (1982) and Anthias and Yuval-Davis (1993) for further elaborations of both race and nationality as 'imagined community'.

3 This emergence of the gene as popular vernacular is evident in the established, pervasive trope of gene-discovery stories (from gay genes to genes for anorexia), transgenic animal inventions (from Dolly the cloned sheep to transgenic pigs for transplant organs) and feats of forensic justice achieved through DNA fingerprinting. Similarly, genetic science is an entrenched motif across genres from science fiction to police procedural. Interestingly, the X-files set out a number of the most potent cultural tropes concerning 'mutant' genes as a metaphor of 'alienness' and its narrative problematisations, in that context, of notions of nation, identity, science and 'truth'.

This chapter elaborates on the themes raised in the previous chapters, particularly focusing on the articulation of genes, race and nation. It takes as its case study, the highly prestigious 1991 Reith Lectures – entitled 'The Language of the Genes' – delivered by Steve Jones, then Reader (soon thereafter, Professor) in Genetics at University College London. The Lectures were formative in a number of respects, not least in its launching of Jones himself as a media personality and cross-over author. They also provide a foreground to uneasy motifs that would become embedded in popular debates concerning the gene – its connections to the ignominious history of colonialism, racial science and its culmination in the holocaust, and its potentialities as a corrective to that history. These motifs, as we saw in Chapter 2, punctuated the imagery and imaginary of *Genetics for Beginners,* the arguably more populist rendering that Jones would go on to co-write in 1993. The 1991 Lectures, delivered by radio, and published subsequently as a quality trade book, also represented a formal public introduction, the first, self-conscious public launch of the recombinant genetic era in the UK. This marked a watershed not only for the particular convergence it represented between science and culture, but for what it suggested about science as a site and source of cultural authorship.

In what follows, I will consider the textual economies, narrative and metaphoric, through which Jones translates the science of genetics. I am particularly interested in his use of the metaphor of 'language' for the ways in which it would seem to democratise, to open up the expert conceptual terrains of science and to invite a familiar, interpretive mode of address to the non-scientific reader. Additionally, the languages of literacy – of reading and writing genes; the conceptualisation of genes as reproductive and metamorphic bodies; and the narrative conventions through which genes are envisioned, on the one hand, as the dominant discursive constituents of 'race' and 'nation', while on the other, are disclaimed for racism and nationalism – will be semiotically traced in the imagined pasts and futures that constitute Jones' vision of genetic science.

I will begin, firstly, with a brief examination of the British Broadcasting Corporation, the immediate context for the Reith Lectures slot, and will consider the role of both in the production of a particular, if contradictory, version of British nation. Against this backdrop, Jones' lectures, with particular emphasis on the first two lectures ('A Message from our Ancestors' and 'The Rules of the Game'), will be considered for the ways which these notions of Britishness mediate his representations of genes, of genetic science, and of Jones himself as a geneticist. Discussion in this context will, as noted above, focus in particular on metaphors of language and literacy. Finally, the analysis will turn to the textual and subtextual motifs surrounding 'race' and 'nation' as articulated through Jones' claims for a redemptive, anti-racist genetics.

The 'Best of British': British Broadcasting, the Reith Lectures and the Production of Nation

Steve Jones' Reith lectures were broadcast over BBC Radio 4 over a five week period in the autumn of 1991.[4] The context of Jones' lectures has particular pertinence for the meanings accruing to them. As part of a tradition of such talks, going back to the inception of the BBC, both Jones' performance as Reith Lecturer and the subject-matter of his lectures necessarily draw on and reinvest in the historical currency of British Broadcasting with its post-war, post-Victorian (and incipiently post-imperial) agenda of, on the one hand, paternalistic moral uplift and public service (the education of the poor towards its 'better British self'), and on the other, of public interest and social responsibility (the promotion of a more democratic society).[5]

The Talks tradition, beginning with the National Lectures in 1928, later becoming the Reith Lectures, perhaps best encapsulate the 'high culture' version of Britishness espoused in the British Broadcasting tradition. The Talks format, and the National Lectures in particular, were, as Scannell and Cardiff (1981) write, 'designed to hold the blue ribbon of broadcasting and to provide, on two or three occasions in the year, for the discussion of issues of major importance and the interpretation of new knowledge by men of distinction in the world of scholarship and affairs' (p. 182). The classed and gendered (and implicitly racialised) paternalism implied in the National Lectures remit where professional intellectuals, particularly those of a liberal persuasion, bring high culture to the public, formed the direct model for the Reith Lectures[6] (and its long history of almost exclusively white, male and middle class lecturers).[7]

---

4 The lectures were broadcast on BBC Radio 4, Wednesday evenings from 13 November–18 December 1991 and repeated on BBC Radio 3 on Saturdays, 16 November–21 December 1991.

5 For further discussion see: Scannell and Cardiff (1981); Hall (1982); MacCabe and Stewart (1986); Scannell (1990).

6 They also set the tone for other talks slots, both on radio and television. For example, the 'David Dimbleby Lecture' broadcast on television in November 1997 follows in this tradition.

7 Indeed, the centrality of both liberalism and Britishness to the slot emerged pointedly in the minor furore, especially pursued the British tabloid, the *Daily Mail*, that surrounded the invitation of Professor Patricia J. Williams to perform as Reith Lecturer in 1997. That Williams was clearly perceived as neither a 'fit' nor 'fitting' choice, accrued not only to the drawing room masculinity and whiteness of the Reithian heritage but also to the liberal-imperial politics which have historically characterised previous lecturers. Williams was not only an African American woman, but an American lawyer, a prominent feminist and a highly regarded theorist of racism culture and the law. See Williams (1997).

With Jones' 'The Language of the Genes', the particular connotations of Britishness and intellectual elitism invested in the Reith Lectures, and in the Reithian tradition more broadly, are thus implicitly constitutive both of Jones' own ostensible profile as the educated 'better self' of Britain and that of the object of his expertise – genetics. In this context, Jones' Lectures were a key moment of transition for both the science and the scientist in a number of respects. Firstly, the Lectures provided a moment where what had been a relatively arcane professional discourse was (re)articulated as a popular narration and where a seemingly unassuming man of science became, of sorts, a star. Jones' own self-effacing presentation as a rather eccentric snail geneticist[8] who sat all day watching snails or sequencing proteins seemed resonant of a particular trope of scientific masculinity: quaint, unassuming and yet genius, a safe, even charmingly gauche, British boffin. Secondly, as science repackaged for a popular (if elite) audience, Jones' translation of the arcane languages and practices of genetics promised both access to and popular ownership of the privileged terrain of bodies (of knowledge and of knowers) that matter.[9] In this context, the constitution of preferred audience was clearly aspirationally located in middle (perhaps even upper-middle) class, professional England. Finally, the resonances of these particular versions of Britishness subliminally associated genetic science with the putative qualities of legitimacy, social responsibility, moral and educative uplift and honourable paternalism connoted by the Reithian tradition itself. Genetics, in other words, was, by this means, conscripted into a mutually validating project of nation (for example, as a national resource) and expertise.

Received Pronunciations: Genes and the Language of Language

[In] its programmes and policies, [the BBC] set out to *address* the nation it had so constructed and then become its "Voice" ... The whole gamut of "national voices" was reflected back to the nation through the medium of the sound waves. Yet the Standard Voice – the "received" accent, pronunciation, tonal pitch of the "BBC voice" – circumscribed and *placed* them all. This was *not*, of course, "Cockney" or "Scouse" or even, quite, "Oxbridge". It was a variant synthesis of the educated, middle-class speech of the Home Counties. (Hall 1982, p. 33, original emphasis)

---

8 In an interview on BBC Radio 4 the week preceding the delivery of his first lecture, Jones made jokes about a life of snail-watching.

9 I have drawn here on the multiple meanings suggested by Judith Butler (1993) in her book of the same name.

Given the historical (indeed almost cliché) centrality of 'received pronunciation'[10] to the BBC's construction of Britishness, the language and literacy metaphors that permeate Jones' interpretive framing of genes take on a particular resonance. Genes become at once a familiar touchstone through which the audiences may aspirationally identify themselves in the articulations of expert culture and agency and an ostensible invitation into the democratisation of science such metaphors seem to herald.

Genes and the Metaphor(s) of Language

As the title of Lecture Series suggests, the notion of genes *as* a language constitutes the framing metaphor for Jones' consideration of the cultural significance of genetic science. In the very first lecture, entitled 'A Message from Our Ancestors', there is a direct attribution both linguistic and narrative functions to genes:

> Genetics is a language, a set of instructions passed from generation to generation. It has a vocabulary – the genes themselves; a grammar – the way in which the inherited information is arranged; and a literature – the thousands of instructions needed to make a human being. (p. 4)

> Our understanding about our place in nature has been transformed by the new ability to read inherited messages from the past … . (p. 2)

> We can use [genes] to piece together a picture of human history more complete than from any other source. (p. 3)

The gene thus emerges as a structural singularity, an embodied inscription of both meaning in the historical sense and function in the material sense. The ascription of a narrative property to the gene is implicit in the language metaphor (and graphic in the history metaphor) with their connotations of recording, of purposeful communication ('messages' are 'passed'; 'instructions' are 'pictures more complete than from any other source'). In this formulation, the genetic scientist as 'reader' decodes, but significantly does not produce; is a discoverer rather than mediator of meanings construed as already embedded, already intact in the structure of human (and, by implication, other species')

---

10 Interestingly, it can be argued that the standardised BBC accent has had something of a demise in recent years, with regional accents (though clearly biased toward the Southern, middle class) somewhat more in evidence. Whether this represents a significant challenge to earlier versions of national identity projected through a particular accent is, however, debatable.

biology. A number of interesting implications accrue to this notion of genes. First is the double movement of the notion of a language of nature. On the one hand, genes are constituted in terms of authorship – as (natural) agentic entities engaged in the purposeful work of meaning production. At the same time, the phenomenon of language itself is essentialised in structuralist, positivistic terms. The denotative properties of *genes as words* – their 'claims' as it were – emerge, naturalised, as equivalent certainties. Indeed, these 'claims' include 'errors in the message, genetic abnormalities' which are taken to represent 'sometimes ... the only clues of shared descent' (p. 3).

Genes and the Metaphor of Text

A second and similar set of implications accrue to the construction of genes as *texts*. As suggested in an earlier passage quoted above, one of the guiding claims of Jones' Lectures is that genes encode an historical record that is more accurate than from any other source. Indeed, at various points throughout his Lectures, Jones dismisses the fields of psychology, education and history, for their purportedly incomplete, inaccurate or ill-intentioned explanations of identities, societies and migrations. In his discussion of scientific racisms and the eugenics movement in the second Lecture *'The Rules of the Game'*, for example, Jones begins by noting the role of biology as a science in the service of prejudice (p. 45), but then quickly reattributes this history as a failing of anthropology (and at other points, psychology and education), a field which 'waited years in trying to sort out divisions into which people could be classified [thus illustrating that] it is only a tiny step from classifying people to judging them' (p. 47). Biology, however, is recuperated with the assertion that it broke 'the ties between the genetics and politics of race' (p. 48). How anthropology's misguided taxonomy of 'imaginary pure races' (p. 47) can be distinguished from contemporary genetic taxonomies, Jones does not explain. One is left with the implication (and at times the explicit assertion, as I discuss further below) that contemporary genetics, because of its unprecedented accuracy, is an antidote to the excesses that accrued to its *not-really-scientific* past.

In this context, the gene appears to represent a peculiar elision of narrative and nature. On the one hand, history (as a constellation of social-cultural practices) is understood literally to *write* genes. Yet at the same time, genes are taken not only to materially *inscribe* historical events, but indeed as superordinate *narrational* records of culture, identity and meaning. As Jones sets out in his introduction,[11] *The Fingerprints of History*:

---

11 Quotes throughout are taken from the book *The Language of the Genes*, which provides a transcription of the original Lectures.

Sometimes history itself is a clue as to where to start. Alex Haley, in his book *Roots*, used documents on the slave trade to try to find his African ancestors. He found only one, Kunta Kinte by name, who had been taken as a slave from the Gambia in 1767. The patterns of genetic diversity in today's black Americans could have told him much more … Alex Haley, by comparing his genes with those from different African countries, might have learned much more about his ancestors than he could from the written records. (p. 7–8)

Jones' invocation of *Roots* as illustrative of the limits of social history and potentiality of genetics takes the textual metaphor to a number of disturbing conclusions. First is that genes inscribe a precise taxonomy of racial-ethnic origins. Here racial, ethnic and national identities elide, emerging as definitive homogeneities – i.e. the putative Gambian gene. Second is the positivist notion that genetic profiling not only (accurately) traces racial, ethnic and national migrations, but that these tracings are meaningful in the larger cultural sense; indeed Jones suggests that they are more meaningful, when they are stripped of political and economic context (that characterise, for example, documentations of the slave trade). Thus, while historical records may suggest the locus of significant genes, the 'real', objective and unadulterated story resides in biology. Historians (like anthropologists) produce imagined communities and identities, while geneticists discern their truths. Here we have the apotheosis of reductive history, scientifically rendered. A DNA map could have saved Alex Haley the trip.

This positivist investment in the precision of genetic profiling, premised as it is on a breathtaking denial of interpretive agency on the part of the geneticist, seems utterly to dismiss historical contingency, ambiguity and complexity. As with the *Roots* example, most of the lectures are preoccupied with migrations, but only as movements of populations, with the political exigencies of those movements (invasion, colonial occupation and underdevelopment, civil war) euphemised or ignored, and with ambiguities of parentage (let alone lineage) entirely unconsidered. Thus the 'roots' and routes of nation and identity become stripped down to discrete, reified biological traits constituted as emblematic of such differences: the 'Hapsburg lip' of European Royalty (p. 3); Gambian DNA that codes for 'sickle-cell' (p. 8); or Kenyans' long legs (p. 23). In textual terms, the 'reading' of genes as documentary evidence of *social* history elides cause and effect, context and biology, both in and as a sweep of nature.

Genes and the Metaphor(s) of Literacy

Literacy – the ability to read and write – connotatively evokes, I would suggest, the chief values associated with a democratic society: freedom of expression, self- and communal-empowerment, and full citizenship. These connotations accrue in no small part from the historical centrality of education to both progressive liberal and revolutionary movements for social liberation. Struggles for education have been central across histories and locations in struggles for freedom, for citizenship, for equality.[12] Similarly, as many commentators have argued,[13] the social relations and social divisions of expertise revolve, in part, around the politics of access to the languages and texts of professionals. The marginalisation of women in (or their exclusion from) scientific professions, for example, has been underpinned by the marginalisation or exclusion of girls and women from science education (Whitelegg 1992) and a constellation of (classed, gendered, racialised and related) practices that effect professional closure through exclusion or selective inclusion (Witz 1992).[14] Finally, literacy and the lack thereof are deeply embedded in the constitution of national identity in at least two key respects: first, as a condition of access to, or marginalisation from, rights and citizenship; and second, as a matrix through which the notional histories, boundaries and exclusionary unities of nationhood are written, read and materialised (Butler 1993).[15] The authorship of nation, the constitution of its preferred memberships (and the preferred readings thereof), are thus embedded in unequal conditions of access to the hegemonic (very often expert) languages – the literal and figurative passwords – of legitimate(d) national identity.

On the one hand, Jones' framing genes and genetics through metaphors of language, text, reading and writing can be said to invoke the liberatory connotations of literacy. Indeed, resonant with the Reithian educational remit, the Lectures represent a not insignificant effort to open the linguistic borders between science and 'ordinary people'. It is not often, after all, that scientists take on (or even have opportunities to take on) an explicitly educative role through popular culture, or engage in professional accountability to those outside (or a least at some distance from) their circles. At the same time, the use of the languages of literacy in this context raise significant questions about the extent to which they represent a substantive challenge to the exclusionary relations of expertise. When Jones tells us what he reads of genes, in other words, he does not confer upon us the ability to read them ourselves. The putative 'language of the genes', notwithstanding its conscription into familiar

---

12 See, for example, Friere (1970); hooks (1982); King Jr. (Washington ed.) (1986).

13 See, for example, Freidson (1970); McNeil (1987); Witz (1992).

14 Margaret Lowe Benston (1992) argues that patriarchal relations of technology accrue from and reinvest in the lack and denial of literacy to women in the languages of technology.

15 I refer here to the triple meaning of Judith Butler's use of the term 'matter': matter as both noun and transitive verb – that which is material, that which is *made* material or *materialised*, and that which *matters*.

analogies, metaphors or popular mythologies, is premised on the language of *genetics* – an expert discourse; a terrain of conceptual authority, complexity and empirical application effectively closed to those who have won no effectual mode or legitimate right of entry. The notion of 'reading genes' is a metaphor which both refers to and obscures the mechanics of the 'reading' process – the constellation of biochemical, biological, digital-technological practices that protocols of recombinant DNA science. However user-friendly it may sound, 'reading' genes is not like reading.

Similarly, the metaphor of writing, although taken up only marginally in the Lectures, is nonetheless implicitly linked to a notion of reading. In this context, the power to 'read' genes is, in part, embedded in the power to manipulate them; in effect to 're-write' the organism – the substantive meaning of *recombinant* DNA techniques.[16] Here too, *editorial* decisions about which genes are meaningful; which genes can or should be mapped, cut, copied or pasted; and which should be deleted, are predicated on the inequalities and dependencies that shore up the boundaries of professional expertise, authority and authorship. The effective closures of the genetics reading/writing community have particular implications for Jones' claims for socio- or historo-genetics. For example, even as the work of professional genetics is embedded in its exclusivity, so too is the historical record ostensibly produced through the 'reading' of genes.[17] Genetics rarefies rather than democratises the journey for 'roots', fostering widening public dependency on a much reduced pool of professional readers and heightening the inequalities that constitute scientific expertise and scientific professions. In its effective re-investment in universalising explanations of social formations and historical movements *as nature*, moreover, it reduces rather than expands what is constituted as meaningful and who may be considered legitimate meaning-makers.

---

16 The creation of hybrid organisms, the 'supermouse' (a mouse with the 'gene for' human growth hormone), for example, or the geep (a genetically engineered combination of sheep and goat), are high profile examples of 're-writing' practices carried out by genetic researchers. However, the biochemistry applied to disaggregate the genetic material of a cell, to clone particular material in culture and to isolate functions can also be construed as an *editorial* process combining the intertwined protocols of 'reading' and 'writing' as active *recombinant* interventions.

17 This is not to suggest that the authorship of written histories is not also boundaried by relations of literacy and expertise. Nonetheless, there is a potential, however latent, to democratise the conventions of readership and authorship, which patently does not accrue to recombinant genetics and the authorial and readerly practices that constitutes contemporary genetic 'literacy'.

Articulated Tensions: Reading Genes – Writing (Anti-)Racism

The tensions in Jones' Lectures surrounding the representational economies of language and literacy inform a similarly ambivalent evaluation of the relationship of genes to questions of race, racism and nation. As I shall discuss below, two rather contradictory currents shape the Jones' Lectures in these respects, as he proposes, on the one hand, a genetics that mitigates against racism and, on the other, a science defined by its impetus to produce modes of racial and national taxonomisation.

Re-mapping Anatomical Geographies

As set out in the second Lecture, Jones proposes genetics as antidote to racism in two key respects, both of which, he argues, represent significant breaks from the past 'grim' relationship between science and racism (Jones 1991, p. 46). First is the argument that genetics challenges and puts paid to the notion of biologically discrete races, a notion which Jones views as the basis of racism:

> Genetics has at last given us a way of testing the pure race theory. (p. 49)

> Individuals – not nations and not races – are the main repository for genes whose function is known. The idea that humanity is divided up into a series of distinct groups is quite wrong. The ancient private homeland in the Caucasus – the cradle of the white race – was just a myth. (Jones 1991, p. 51)

> Even forty years ago, racial stereotypes of the most predictable kind were the norm among psychologists. They were the last remnant of the idea of racial type, a view which biologists had abandoned long before. (p. 52)

Science, in this construction, emerges ambivalently: on the one hand as implicated in (though perhaps not as prominently as in, for example, less centrally scientific fields like psychology) such commonsenses of racial difference; and yet, on the other hand, as a corrective discourse which now, thanks to developments in genetics, has the tools to reveal the erroneous and unscientific foundations of racial discrimination. In this context, Jones points out that 'race' itself is an unstable category with historically shifting notions of the boundaries understood to constitute a racial identity or characteristic. Jones points up skin colour in particular as an historically contested marker of race (p. 46). Genetics, he argues, empirically destabilises racial categories, having the capacity both to reveal *individuals* rather than groups as the repository of biological difference(s) and to thus disassociate (the taxonomisation of) *traits* from (the construction of) identity. This, in turn, has implications for the relationship between 'race' and racism:

Other creatures vary much more from place to place [than humans] ... The genetic differences between the snail populations of two Pyrenean valleys is much greater than that between Australian aboriginals and ourselves. The difference between the highland and the lowland populations of the mountain gorilla a few miles apart in central Africa is more than that between any two human groups. If you are a snail or a mountain gorilla, it makes good biological sense to be a racist; but if you're not, you have to accept the fact that humans are a tediously uniform species. (Jones 1991, p. 51)

Although this passage is written in a rather tongue-in-cheek vein, there are serious implications. First is an understanding that racism is a consequence (logical, natural) of racial difference. Second is that particular 'facts' of racial difference can be 'proven' (or disproven) empirically. In this understanding, racism is wrong when or because there is not enough 'real' difference to justify it. Genetics offers a tool to remedy the injustices caused by *illegitimate* racial categorisation, as opposed to taxonomic categorisation *per se*. There is clearly an ambivalent (though perhaps unintended) message here; for if genetics reveals *authentic* differences and these differences can be taxonomically validated, then what is implied is a not a rejection of 'race', but its reconstitution – the grounds, in other words, for a 'new racism'.[18]

Thus, at the same time that Jones appears to reject 'race' as a meaningful category, it re-emerges, threading through the text in an uneasy counter-motif. Race, for example, is a key analogy through which Jones explains the mechanisms of genetic inheritance.[19] References, as noted earlier, to royal

---

18 Martin Barker (1981) made a comparable point about the 'new racism' of sociobiology and ethology, sciences which have eschewed 'dated' and conventional languages of racial hierarchy in favour of a discourse of 'ways of life'; human tendencies to form exclusive and biological dispositions towards xenophobia. Jones, in common with the assumptions of these new sociobiological and ethological discourses, explains racism chiefly as a logical, natural consequence of racial difference in and of itself.

19 In explaining the mechanisms of matrilineal inheritance Jones writes: '[s]perm contribute very few mitochondria when they fertilise an egg so, like Jewishness, this DNA is passed through the female line. It contains the history of the world's women, with almost no male interference' (p. 8). This is a characteristic elision of biological mechanisms with social history. Jewishness is presented here both as a genetic trait and a closed matrilineal line of descent. This is notwithstanding that Jewish identity can be assumed through conversion. It is hard to see how the social history of Jewish people supports a genetic approach to its definition, or indeed any other racial/ethnic/national classification. Jones' notion that genetics remedies the problems of racial mis-classification–racism would seem to be undermined quite graphically in this example.

pedigrees and to racial- or ethnically-based taxonomies of disease[20] or physical traits recuperate the foundationalist notions of human difference that Jones disclaims for contemporary genetics. So too do characteristic elisions of racial and national identity that permeate the text:

> We, the British, contain more hunting genes than do, say, the Greeks, who had rolled over the earlier economy and absorbed its genes long before. (Jones 1991, p. 41)

> Throughout modern Europe, we can see patches of genes which reflect the successes and failures of nations and economies long gone … today's southern Italians and Sicilians are still genetically distinct from their compatriots in the north … . (Ibid.)

> A genetic map of Europe shows that most language boundaries are in fact regions of genetic change. In Wales, there are genetic differences between Welsh and English speakers … . (Ibid., p. 42)

> Genes persist far longer and can tell us much more about the past [than language] … We see this in the Etruscans [whose language and culture are now extinct]. (Ibid., p. 43)

> We can use genes to make a family tree of human kind, and to reconstruct the relationships of the peoples of the world. Africans as a group are on a branch of the human family which split off from the others rather early on, and most of the rest of us are more closely related to each other than we are to the populations so far tested in Africa. (p. 52)

These passages represent distinctively corporealised, reified and *racialised* understandings of nation. Nation elides with genetic profiles; multicultural constituencies are obscured, mapped out instead as discrete homogeneities (*the* English, *the* Welsh, *the* southern Italians). Indeed, as with the Etruscan example, genes are understood to encode national identity even where its usual cultural markers (e.g. language) are 'extinct'. The representation, furthermore, of Africans as a branch of humanity 'split off' from 'the rest of us' not only connotatively elides Africa as nation (as distinct from continent, or multiple multi-ethnic nations) with African as black race (even as 'English' evokes

---

20 Tay-Sachs, for example, as a European Jewish trait; sickle cell as an African trait and so on.

'white') but it resonates with colonial-eugenic discourses of racial hierarchy.[21] Such implications are reinforced through discussions, elsewhere in the Lectures, of 'modern primitive' cultures like the Yanomamo tribe of South America '[among whom] [r]ape, murder and theft is common' (p. 37) or of India as the repository of uncivilised, hyper-patriarchal values and practices, putative home of sex selection practices where 'being female is often seen as a genetic disease' (p. 58)[22] (but which nevertheless, or indeed because of this, may carry edifying messages about 'ourselves').

Re-reading Genetics after Eugenics

The subtextual (and perhaps unintended) mobilisation of the racist commonsenses of colonial discourse inform similar contradictions in Jones' second, and related, claim for a remedial genetics revolving specifically around the question of eugenics:

Much of the story of the genetics of race – a field promoted by some of the most eminent scientists of their day – turns out to have been prejudice dressed

---

21 At other moments, this discourse is quite graphic, as in Jones' representation, in colonial-anthropological vein, of the Yanomamo:

We can get some idea what life was like by looking at modern tribal peoples ... The Yanomamo Indians of South America ... call themselves "the fierce people", with good reason. The commonest cause of death is violence ... [they] exist in a series of small bands. These are in constant conflict. Rape, murder and theft are common.

Social Systems based on hunting and gathering – as all were for 90% of human history – may have been like this. It is dangerous to make too much of what one tribal culture like the Yanomamo does. Others – such as the Bushmen – are far better behaved. (Jones 1991, pp 36–7)

22 Jones contrasts this with Britain and 'most Western Countries' where he imagines that '[m]ost people would ... see the possibility of terminating a pregnancy just because it is the wrong sex as being ethically unacceptable' (p. 59). While Jones in fact does not reject sex selection on health grounds (and does suggest that 'most people' might be more prepared to accept sex-preselection (separation of x and y sperm rather than abortion of 'wrong' sex foetuses), India is nevertheless invoked as an example of the unacceptable face of sex selection, while its more 'ethically justifiable' use is suggestively located in Britain and 'most Western Countries' (Jones 1991, p. 58). Later in the lecture, Jones invokes the Indian village as an iconographic site of backwardness characterised by remoteness and therefore 'inbreeding'. This he contrasts, in rather utopian vein, to the process of population mixing through which both biological and therefore cultural differences (e.g. between England and Scotland) will be 'even[ed] out' (p. 65), thus minimising the chances of recessive illness. Here Jones again elides genetic and cultural differences; disease and racial/ethnic/national identity.

up as science, a classic example of the way that biology should not be used to help us understand ourselves. Most geneticists are genuinely ashamed of the early history of their subject and make every effort to distance themselves from it. (Ibid., p. 54)

Jones argues that contemporary genetics breaks from its early history in two key respects. First, as previously noted, Jones makes the sustained claim that contemporary genetics no longer deals in imaginary traits, but in empirical facts: '[eugenics'] disgrace was made more complete by simple errors: the genes involved often did not exist outside the doctor's imagination' (Jones 1991, p. 57). This would seem to square uneasily, at the very least, with the association, elsewhere in the lectures, of genetics with notional constructions of racial–national traits, heritage and character. Second, Jones argues that geneticists are no longer interested in the grand project of eugenic social engineering. He states:

> We now have the answers to many of the genetical questions which obsessed the eugenics movement. However, there has been an odd shift in attitude: modern geneticists scarcely involve themselves with what their work implies for the future of humanity. They feel responsible to people rather than to populations, to individuals rather than to posterity'. (Ibid., p. 57)

> No serious scientist now has the slightest interest in reproducing a genetically planned society. (Ibid., p. 58)

It is not clear how the individualisation of genetic selection obviates eugenics, particularly if contemporary genetics has 'given us the answers to many of the genetical questions which obsessed the eugenics movement'. Indeed, it can be argued that an increasingly commonsense ethos of reproductive screening might obviate the need for eugenic masters or grand plans; nor is it clear how a more localised focus for genetic science eschews the quality of 'planning'. Clearly its applications do not operate in an institutional or conceptual vacuum. Jones, moreover, states that 'moral problems about the quality of people and whether we can, or should, make choices based on genes' (ibid.) inevitably accrue from genetic screening. Yet this acknowledgement of a shared institutional ethos between past and present genetics is nevertheless dismissed in a double-edged recuperation of scientific agency. For Jones interprets the 'disgrace' of early genetics as a consequence of the prejudicial, hubristic and unrealistic intentionalities of (bad) scientists. In this context, and in a stunning contrast to Jones' own use of such grand narratives, contemporary scientists are construed no longer to be dangerous utopian visionaries but benign and modest practitioners invested in tracing lost histories and preventing disease, both quests now (properly) dislocated from questions of morality or politics. Jones concludes:

> Fictional Utopias nearly all seem to evolve in roughly the same way. A master race imposes its will on lesser breeds, only to meet its doom because of its own biological failings ... Evolution always builds on its weaknesses, rather than making a fresh start. It is this lack of a grand plan which has made life so adaptable, and humans – the greatest opportunists of all – so successful. (p. 66)

Jones' analogy argues for a genetics redeemed of its past 'failings', a eugenics cleansed of its previous obsessions and grandiosity, a progress and language, at last, restored to nature. Genetics, in other words, relocated to the putatively benign neutrality of natural selection, is thus seen to be rescued from ideology and its attendant bad intentionalities.

Conclusion

> As it became possible to look to "nature" for explanations for human society and character, so too was Darwin's loan "read back", as Strathern describes it, enabling nature to become subject to visions of social improvement. This traffic ... much as it has informed the Euro-American imagination more broadly ... has specific roots in England where the national culture has long been formed in relation to the zig-zagging repeat of analogies linking nature, progress and society. (Franklin 1997, p. 99)

As I have traced throughout this chapter, Jones' use of metaphors of language and literacy and his claims for an anti-racist genetics produce the seductive image of a science harnessing nature in the service of democracy and social progress. There is a distinct (if limited) register of popularisation, an apparently alternative (non-programmatic) programme of liberal social improvement, a promised breaking of links with a dubious past except insofar as lessons that have been learnt to the better. Yet on closer inspection, we find a science proposed as an antidote to problems in which it remains foundationally embedded. The democratising language of the Lectures is undercut, even duplicitous, as it masks and misdirects the institutional closures of expert agency. The claims for an anti-racist genetics, however well-intentioned, emerge as essentially rhetorical, window dressing, a contradiction in terms as the foundationalist notions of race and nation which Jones appears to dismiss are also powerfully reinscribed. Finally, Jones' casting of contemporary genetics as a science no longer tainted by ideology or grandiose intentionalities, is belied by the ideological underpinnings of his own readings of genes and by the power relations, of which he seems not to be aware, accruing to his claims for genetic literacy.

The profound tensions characterising Jones' Lectures raise uncomfortable questions about the limitations and potentialities that might constitute genetic science, whatever version is pursued. Can there be a genetics divorced from its own history? If the conceptualisation of genes is intrinsically embedded in foundationalist epistemology and ontological taxonomies, how is an anti-racist genetics possible? To paraphrase Illich (1976), how can the progressive genetification of life not produce a cultural iatrogenesis that exacerbates existing inequalities or creates unprecedented dependencies? Finally, does a liberal standpoint intrinsically support the totalitarian tendencies of scientific progress even as it promises to recuperate them for a better world?

Chapter 4
Monster X: On Gays, Genes, Mothers and Mad Scientists[1]

As a researcher in a STEM subject I recognise part of the problem. Crudely put, the media look for two types of science story. One is the "isn't this fantastic look what the future holds" type. The other is the "oh no look at what all these crazy scientists have done we are all going to die" type. (stoneturner27, 14 August 2013; 6:48pm, CIF, *Guardian*)[2]

Introduction

Most people will never enter a genetics laboratory or be educated into first hand genetics knowledge or practice. Lay understandings are primarily constituted, second hand, through intermediating languages and sources and, increasingly, through niche encounters in medical (as patients or potential patients), forensic and commercial[3] contexts. These encounters are however part of, and intimately informed by, a super-ordinate media ecology of which reportage is a primary

---

1 An earlier version of this chapter appeared as: 'Pedagogic Panic or Deconstructive Dilemma: Gay Genes in the Popular Press' (1999), in Debbie Epstein and James Sears, *A Dangerous Knowing: Sexual Pedagogies and the "Master" Narrative*. London. Cassell.

2 This excerpted quote is taken from CIF (Comment is Free) following the article 'Why Open Access Isn't Enough in Itself' which concerned the misfit between *Guardian* reportage and the scientific work of the article's author, Ellen Collins. stoneturner27 goes on to argue: 'The media want a story, I can understand that. The reality of most scientific research – and probably social science research as well – is that it is much more nuanced. So your experience doesn't surprise me, and the origin of the problem lies in a journalist's *filtering* of your research to fit in to a particular narrative' (*my emphasis*). While this point is valid, it does beg an interesting question. There is an unspoken implication here that scientific research does not involve 'filtering' or 'fitting a frame'. As I go on to argue in this chapter, if we consider the underpinning assumptions of the gay gene research (e.g. its framing understanding of 'gay' as a unified social-biological identity), such an inference would seem to be pointedly contradicted. This chapter suggests that the 'gay-gene' reportage articulates multiple narrational frames, not all of which only accrue to journalism, and that the arising interest of narrative framing accrues to its particulars as much as to the institutional location of their authors.

3 See, for example, Kramer (2011a; 2011b) for further discussion of commerce (and its mediatisation) in genealogical DNA testing.

constituent; news is a foregrounding site for the public dissemination of science into daily life. Journalism is a distinctive media form, particularly with respect to its capital *vis-à-vis* 'truth' and 'fact'. Balance and objectivity are widely cited, ideal-type journalistic values (even as they are contested terms and contested assessments of the journalistic enterprise). And 'news' is characteristically distinguished from 'opinion', including in leaders about new scientific discoveries. Journalism is also distinctive for its location at the forefront of popular (as opposed to professional or scholarly) discourse.

For the past two decades, at least since the release of the film version of *Jurassic Park* in 1993, mass media has been rapidly and progressively permeated with genetic themes and references. At the same time, as suggested in the preceding chapters, popular cultural idiomatics have informed professional discourse and the scientific imaginary itself, perhaps most notably at particular points where scientists themselves enter into public discourse and take on the function of popularisation, or at moments – as in the case study for this chapter – where scientific claims (or claims about science) collide with, or play out on the terrain of, wider cultural tensions, anxieties or social movements. This chapter explores the *signification spiral* and inflammatory reportage that attended the announced finding in 1993 of a 'gay gene'. Focusing on a range of British newspapers, representing tabloid and broadsheets as well as distinctive political orientations from Conservative to Labour, the chapter traces the contestatory reception of diagnostic genetics on a terrain which was itself in a state of socio-political contestation – gay rights versus gay pathology – and which, consequentially, stood in for and evoked larger cultural tensions concerning reproduction and choice, and the (il)legitimate social reach of science and government (and science *as* government).

Breaking News

On 16 July 1993, the journal *Science* published a report by Dean Hamer and his research team at the National Institutes of Health in the US that they had discovered a genetic basis for homosexuality.[4] The 'gay gene' discovery story was picked up by international media; in the UK, the story broke on the 16th as a broadsheet story in *The Times* and the *Independent*. On 17 July all of the main British papers, broadsheet and tabloid, carried the story (all carried multiple stories). Discussion continued in *The Times* and the *Independent* until the 24th and 19th respectively, while disappearing from the tabloids after the first day. Although breaking the story on the 17th, the *Guardian* carried out most of its reporting on the 'gay gene' discovery in the final week of July. The timing of this announcement had distinctive resonances in the UK context, still reeling

---

4 Hamer et al. (1993). See also Hamer and Copeland (1994).

from the highly controversial passage of Section 28 of the Local Government Act in 1988, which banned the 'promotion of homosexuality' and stigmatised gay families (or families with gay members) as 'pretended'.[5]

Framing Gay Gene Science

Hamer's findings of putative genetic links to homosexuality were based, as Ewing[6] (1995) notes, on a two-part study of 40 pairs of self-identified gay brothers. The first part involved drawing up family trees with the aim of identifying other gay relatives – the result of which appeared to show more gay relatives on the maternal than on the paternal side of the family. Hamer and his team inferred from this a pattern of maternal inheritance (that is, that homosexuality was passed down as a recessive trait on the X chromosome). The second part involved an analysis of the DNA of the gay brothers 'to see if they had inherited genes in common that could, by inference, be linked to their sexuality' (p. 2). It is worth noting that Hamer's investigation into the possible genetic basis of homosexuality grew out of his research on AIDS related cancers, a point which some of the news reports picked up. Thus AIDS was the immediate conceptual frame and referent for both the gene research itself and for its coverage in the popular press. A second and equally significant frame for the research was its premises concerning the meaning of 'gay' as a universal, singular, social cum biological category, in which desire, behaviour and identity, simply and straight-forwardly line up.[7] A corollary assumption appeared to inform the making of family trees for the research subjects, relying on second-hand attributions of heterosexual or gay identities of family members, attributions that may or may not have squared with how those family

---

5 See Stacey (1991). As Stacey notes, the public contestation over Section 28 was considerable.

6 Christine Ewing presented her critical assessment of Hamer's research at a scientific conference and kindly provided me with the transcript. Her discussion drew from her vantage point as a biologist and focused less on the popular discourses surrounding (and provoked by) Hamer's research than on the logics driving the research methodology. The methodological inconsistencies (and non-replicability) of the project in the years that followed this announcement were not widely taken up in scientific circles or in popular discourse. This was with the notable exception of some US Christian groups who argued that the flawed research proved that homosexuality was a (depraved) choice and not natural. This line of argument aimed to counter a USA-based gay rights activist argument that being 'born gay' or 'naturally gay' invalidated homophobic prejudice. This nature-based rights claim was not a position significantly taken up by gay rights activists in the UK. See also Brookey (2002).

7 Implicitly, the study elided lesbian and gay biologies-gay etiologies, as it focused on male subjects but made generalised extrapolations about 'gay' identities.

members understood or defined themselves.[8] Finally, in addition to his reported findings, Hamer concluded his *Science* article with a plea that the research should not be misused to intensify stigmatisation of and discrimination against gay people (p. 5). Some discussion of the research methodology, albeit with varying degrees detail,[9] and acknowledgement of Hamer's *caveat* were also picked up by all the papers.

On the face of it, and as Hamer's team seemed to be uncomfortably aware on evidence of their *caveat*, the notion of a 'gay gene' might seem entirely consistent with the underpinning assumptions of both scientific and commonsense homophobias. Its plausibility, for example, is dependent on the presumption of 'gay' not only as a unified social category, but as a singular, organic problem[10] from which can be inferred biological origins. Both assumptions have underscored sexuality sciences since the nineteenth century where '[h]omosexuals were one of a number of … *internal others* explored for stigmata of degeneracy within the West – alongside criminals, prostitutes, the feebleminded – whose bodies were believed to carry the germs of ruin' (Terry 1997, p. 274).[11] More recently, bio-medical AIDS research has constituted a dramatic site for the reconstitution of such 'languages of risk' in which germs and gays elide as 'deviant', dangerous and deadly bodies.[12] Against this backdrop 'gay genes' inevitably resonate with, indeed seem simply to offer another, this time hereditarian, version of pre-existing venereological commonsenses surrounding homosexuality and (and *as*) sexually transmitted disease.

The 'gay gene' also fits with the dominant trope of 'gene discovery' – the perpetual succession of conditions, biological and social, which are said to have genetic origins. Gene-discovery stories, from cancer to crime, from schizophrenia to anorexia, have become a commonplace feature of both scientific and popular

---

8 A further framing assumption was that a study involving a very small, geographically localised sample, exclusively of men who defined themselves as gay, was generalisable to gay identites and lives as such.

9 Unsurprisingly, the broadsheets had somewhat more elaborated descriptions of the research than the tabloids. Indeed, what did appear in the tabloids generally appeared to be excerpted verbatim from the broadsheet reporting.

10 I refer here also to Lemke's (2004) Foucauldian critique of *problematisation*, as a consequence (and artefact) of the diagnostic genetic gaze, as opposed to the other way around.

11 Terry notes that there have also been resistant strands of sexuality science, for example the sexological approach of Magnus Hirschfield, which have tended to regard homosexuals as 'benign natural anomalies, afflicted not by biological defects but by the social hostility that surrounded them' (ibid.).

12 For further discussion of the ways in which risk discourse historically developed with respect to HIV and AIDS, see, for example, Patton (1985); Watney (1988); Redman (1997).

discourse. Typically it is those conditions or characteristics, already construed as medical or social problems, that become candidates for the seeking of 'candidate genes'.[13] 'Treatment', 'cure', 'prevention', 'solution' are all implicit in the diagnostic framing of both the science and the objects of its enquiry. In this context, the ascription of causality (even tentatively) to genes, aligns with their *problematisation*[14] in the first place – as bodies which require explanation.

The neo-essentialism implicit in genetic science itself has its adjunct in the dominant pathologised 'otherness' that has historically ascribed to lesbians and gays in the popular press.[15] As many commentators[16] have noted, social conservative culture wars surrounding the presence or possibilities of lesbians and gays in schools, making families and forging (organised) claims for civil equality recur regularly, and formed a potent backdrop to pre- and proceeding the 'gay-gene' discovery story. Epstein's (1997) analysis of the myriad antecedents to the 1994 Jane Brown panic provides a salient illustration.[17] If such notions constitute the dominant conceptual field in which an imperative search for 'gay genes' was imaginable in the first place, its putative 'discovery' would seem only to further entrench them.

Given the ways in which 'gay genes' seemed to embody the homophobic trajectories that have broadly characterised both sexuality science and popular reportage, one might therefore expect that the press treatment of the 'gay gene' discovery would have inevitably followed in similar suit – just another homosexual monster story. Yet, as we shall see, when 'gay genes' hit the popular press, a replay of such familiar ground is what *did not happen*. Indeed, if anything, the 'gay gene' stories marked not only a moment of *departure from*, but a significant *destabilisation of*, familiar phobic narratives of homosexuality.

In what follows, I will attempt to account for the surprising displacement of these narratives. Through a close examination of British reportage, I will trace the ways in which the 'gay gene' stories marked a point of convergence for a number of discourses, in particular those of science, homosexuality and abortion,

---

13 When a condition is assumed, in the first instance, to have sole or partial genetic origins, the term 'candidate gene' is used to refer to the particular genetic material suspected to confirm that hypothesis.

14 See Chapter 1 of this book for extended discussion of Lemke's (2004) argument concerning *problematisation* as a driver, as well as an effect, of genetic diagnostic logic.

15 This is notwithstanding that the climate surrounding LGBT lives has significantly liberalised (and consequently de-pathologised) in some contexts.

16 See, for example, Stacey (1991); Epstein (1993).

17 Jane Brown's refusal of subsidised (though still expensive) tickets to a ballet of *Romeo and Juliet* in the summer of 1993 hit the headlines in January 1994, developing into a protracted media hysteria of homophobic panic. Epstein traces the interplay of (anti-)homophobic and (anti-)racist narratives deployed in the Jane Brown reportage.

whose interplay served to rupture dominant homophobic commonsense (even as it summoned into the fore other, arguably more potent, folk devils).

On the Horns of a Moral Dilemma: (Im)Partial Science?

If the 'master narratives' of homosexuality are predominantly framed through the horror genre, those of science are rather more ambivalent. Press take-up and treatment of scientific discourse characteristically occupy three competing narrative conventions. Perhaps the most salient and familiar of these constitutes science as reservoir of natural facts; an objective, transcendent endeavour beyond politics and cultural meaning; a society of experts and body of expertise whose due is deference and celebration; a sweeping tide of progress, intrinsically progressive, rushing toward Disney's Tomorrowland in all its incipient wonder.

Yet science also has its monster narratives, coexisting contrarily, and yet in obstinate synchronicity with the presumptions of its objectivity and rational benevolence. The notion of science in the service of an evil agenda, for example, is deployed in the various genres of 'Hitler horror', with their spectres of human experimentation, genocidal apocalypse, reproductive slavery, clones and (sub)human–animal hybrids. As a reference point or spectacle, such dystopian mythologies bespeak both an elaborated cultural recognition of, and latent anxieties about, the powers of contemporary science, its inaccessible languages, its privilege and its grasp over life and death. Yet if the figure of the mad, bad scientist embodies such anxieties, it also, through its baroque grotesquerie, neutralises them. Against such a yardstick, science-as-usual can appear banalised, easily recuperated, reconfirmed.

The interplay of these narratives emerge graphically in the ambivalent rumblings of celebration and disquiet that have greeted contemporary genetics. We are now accustomed to the 'discoveries' of an ever widening array of genes. Yet if heroic promises of rescue, cure and resurrection seem embodied in the mastery of molecular space, so too do dispossessions and annihilation. Indeed, the gene has embedded in popular vernacular as the quintessential metaphor of origins, the protein (and protean) medium of destiny, a higher order as marvellously mundane as an architectural blueprint, and yet a weapon, in the wrong hands of course, of mass destruction.

When genetic discoveries seem to bespeak laudable motives, the prevention of 'serious' disease, for example, the science goes largely unproblematised. Announcements are made, embellished with wishful extrapolations about treatments and cures for the future, the only *caveat* that this is only an early stage, more work needs to be done. The reported 'discoveries' of oncogenes,[18] genes

---

18 'Oncogenes' are understood to cause, or to be implicated causally in cancer.

for cystic fibrosis, Duchenne Muscular Dystrophy, anorexia and Huntington's chorea, for example, are classic examples. Indeed, just this morning, as I was sitting down to breakfast, the radio announced the 'successful' genetic engineering of mice with three times the normal muscle mass. This research, it was asserted, would lead to greater understanding and possible cure of muscular dystrophy. End of story.

On the flip side, it is no accident that of all the new reproductive technologies on the horizon, it has been the prospects of human cloning; of the interbreeding of human and animal; and of germ-line experimentation of human embryos that have generated some of the greatest public consternation. Not only do these practices seem to most fit the clichéd iconography of fallen science/scientist, but their perceived extremity produces a permissive space in which the powerful can momentarily be identified as 'other', the mighty brought low, the insider revealed as outcast. The fears invoked in the moral panic surrounding the cloning of the sheep 'Dolly' played precisely to the conventions of a modern Frankenstein story.[19] It was the scientist, and the hubris his creation embodied, who was taken to presage disaster. The quaint appellation of 'Dolly' provided the point of identification, for an innocent and vulnerable 'us', the potential human objects of the morbid, metamorphic fantasies of the scientist-who-would-be-god.

More often than not, however, it is a third narrative frame, the *moral dilemma*, rather than the moral panic, which greets the perceived slippages of science from its moral high-ground. The 'moral dilemma' posits that while science itself is a neutral, objective practice, it nonetheless, in particular instances, generates difficult ethical questions. These questions typically focus on the uses and potential abuses of the science. Fears tend to be displaced on to other, 'biased' interests – commercial, governmental or 'society', for example – rather than those of scientists themselves. The 'moral dilemma' is, in this sense, a narrative frame in which both the benevolent and monster narratives of science almost seamlessly coexist, even as the former is delicately questioned, and the latter is apparently disclaimed.

Mapping Moral Minefields

Broadsheet headlines marking the discovery of 'gay genes' explicitly framed the story as a 'moral dilemma':

Gay Gene Raises Host of Issues. (Connor, *Independent*, 16 July 1993)

"Gay gene" Raises Screening Fear. (Hawkes, *The Times*, 17 July 1993)

---

19 For extended discussion, see Franklin (2007).

Scientists Claimed this week that Homosexuality in men is Influenced by Genes. But who does the Discovery help? Tim Radford Reports on the Moral Minefield of Genetics. (Radford, *Guardian*, 17 July 1993)[20]

The tabloids also announced the 'gay gene' discovery with notable ambivalence. The *Daily Mail* headline read: Gay Genes and a Moral Minefield (Jones, 17 July 1993). The *Sun* (Watson, 17 July 1993) juxtaposed three bold-print headlines for the same story:

Mums Pass Gay Gene to Sons say Doctors. (main headline)

Parents may demand abortions after tests. (sub-headline)

Don't try to eradicate us. (bottom-line, credited to Gay Rights Group Stonewall)

The *Daily Mirror* (Swain, 17 July 1993) similarly juxtaposed:

Battle Looms over Scientific Discovery. (top of the page)

Men inherit Gay Genes from Mum. (main headline)

Inherited link is proved. (inset mid-story)

Seek out and Destroy Fears. (inset sub-story headline)

Encapsulated in these headlines are a number of contestations about science – its rational precision, its claims to political neutrality, its (disinterested) moral agency.

Tentative Truths

Ambivalence around the notion for example that there *is* a 'gay gene', that scientists 'proved' the link, emerged in both the tabloids and the broadsheets, albeit in quite distinct ways. In the tabloids, as suggested by the *Mirror* inset headline, the 'fact' of gay genes seemed to be taken as a given. Indeed, as Steve Jones[21] himself asserted (but did not explain) in his article for the *Daily Mail* (Jones 17 July 1993):

---

20 This appeared in bold print over the main headline 'Your Mother Should Know'.

21 Steve Jones emerged as perhaps the foremost 'popular' geneticist in Britain in 1991, when he gave the Reith Lecture series on BBC Radio 4. Since then, he has been involved in various forms of popularisation/popular education around genetics including on television, the popular press, radio and in books intended for a lay-readership (including *Genetics For Beginners*, Jones and Van Loon (1993)).

> There have been many earlier claims of a "gay gene". All were rubbish. This one
> almost certainly is not. The gene has been tracked to a short section of DNA.
> Although there are hundreds of genes hidden in there, the one for homosexual
> behaviour is probably among them.

The literal text of the passage is, in fact, an optimistic rather than absolute
reading of the factual existence of 'gay genes'. Yet, Jones' own media prominence
as genetics expert would seem to render the 'probably' and 'almost certainly',
juxtaposed with the 'all were rubbish' of prior claims, as more a statement of
scientific modesty than one of doubt. In both the *Mirror* and the *Sun* very brief
summaries of Hamer et al.'s research methods were presented without comment
(or explanation), the 'dilemma' instead as I shall discuss further below, centred
on the implications of the 'find'.

The broadsheets, by contrast, provided much more pointed equivocation
both on 'gay genes' as facts and genetic science as a stable body of expertise.
Terms such as 'precisely located' or 'pin-pointed' juxtaposed against 'theory'
or 'claimed' suggested a more resistant relationship to the science. In an article
appearing in the *Independent* for example, reporters Connor and Wilkie state:

> Scientists can now locate the *precise position* of a gene and find out what it is
> responsible for, and what variations of that gene exist. (Connor and Wilkie, 18
> July 1993, emphasis mine)

Yet, later in that same article, 'gay genes' appear to be disclaimed from this
narrative:

> It is easier to say what has not been discovered: the researchers in the United
> States whose findings were announced last week have *not* found a gene that
> causes homosexuality and they have *not* proved that homosexuality is heredity.
> They believe they have evidence linking a region of the X chromosome, which
> men inherit from their mothers – with the sexual orientation of some gay men
> … . (Ibid., original emphasis)

These passages are significant for their validation of the rational precision
of genetic science on the one hand and, on the other, the instability of the
'gay gene' as a *particular* fact. At the same time, the attribution of 'they believe'
suggests, as did Jones' comment discussed above, that grandiosity is not so
much a product of scientific claims, but of their misinterpretation (for which
this article is possibly intended as a corrective).

Yet both the *Independent* and *The Times*, in addition to providing considerably
more detail about Hamer et al.'s research methods, also raised tentative questions
about their validity:

Research has found a common genetic pattern in 33 pairs of homosexual brothers, but other scientists say the evidence so far is only inferential and *the statistical basis of the study is weak*. Evidence has been growing that there is an inherited component in homosexuality. It occurs more often, for example, in identical twins … . (Hawkes, *The Times*, 16 July 1993, my emphasis)

The scientists do not know why seven of the 40 pairs of gay brothers do not appear to have the same genetic markers. Dr Hamer said these gay men may have inherited other genes that are associated with homosexuality or they might be influenced by environmental factors or life experiences. (Connor, *Independent*, 16 July 1993)

In both of these passages, however, the caveats about the weaknesses of the study are immediately recuperated through the flow of the article: in *The Times* by an abrupt, non-sequitur (and unelaborated) reassertion of the evidentiary basis for the claim of a 'genetic component to homosexuality'; and in the *Independent* by an alternative genetics-led explanation of homosexuality in the brothers who lacked the expected 'markers'.[22]

The fundamental premise of the claim of a 'gay gene', i.e. that 'gay' can be treated as a 'factually' unified identity category, was also questioned – albeit in only one editorial, again in the *Independent*:

Um, but it's absurd isn't it? This notion that homosexuality, or the predisposition thereto, has been genetically *located* … For a start, homosexuality is an abstract noun, a cultural construct with a short historical life, as was famously pointed out by Foucault …

Genes decide this, Genes decide that. Genes affect this. Genes affect that. It would be extremely odd if they didn't affect sexuality, since that would rather imply that our sexuality had no basis in our physical being. But it is cultures that decide what constitutes homosexual behaviour or a homosexual disposition. Indeed, one could argue that our own culture has yet to make its own decision on this point. (Fenton, *Independent*, 19 July 1993, original emphasis)

Here the perceived absurdity of a universalised (and trans-historical) *inference* of 'gay genes' underpinned a radical destabilisation of the guiding assumptions about the truth values of genetic science and expertise that otherwise, at least partially, constituted the 'gay gene' reportage. In the balance of broadsheet reporting, however, the 'dilemma' about scientific accuracy was framed more

---

22 It might be added, that seven seems a rather significant number in a sample of 40, a point which was implied in *The Times* article, but not explicitly noted.

tentatively, and relegated to the background of more urgently deployed concerns about the moral status and potential (mis)use of the science.

A Balanced Morality

Despite the fact that it may rarely be carried out in practice, the notion of *balance* is central to commonsense claims of both journalistic and scientific integrity. The distinction between broadsheets and tabloids as high and low culture respectively rests largely on the extent to which they are taken to participate (or not) in the liberal rationality of *representing both/all sides*. The very framing of the 'gay gene' stories as a moral dilemma would seem to signal such balance, indeed to function as a recuperative metaphor for scientificity itself. Thus the interests of journalistic balance were contiguous both with concerns about a possibly unbalanced (i.e. biased) science and with an imperative to recover a middle (i.e. neutral) ground:

> Steve Jones, professor of genetics at University College, London, said "What I overwhelmingly hope for is that this research will not be used to make moral judgements. The findings are scientifically fascinating, but socially irrelevant". (Connor and Whitfield, *Independent*, 17 July 1993)

> Discovering that homosexuality may have a genetic component tells us nothing about the moral or social status of homosexuality. If we choose to classify homosexuality as a disease, we have made a moral choice: it does not follow from the science of molecular genetics. There are no values to be discovered in the double helix of DNA. (Connor and Wilkie, *Independent*, 18 July 1993)

As these passages suggest, the association of genes not only with homosexuality, but specifically with *homophobia*, problematised the notion that the science was morally neutral. The impetus to recuperate scientific integrity seemed to require several rather paradoxical disassociations. For example, the notion that 'the findings are scientifically fascinating, but socially irrelevant' suggests that in a science of homosexuality, the science can nonetheless be disaggregated from the object of its scrutiny; the meaning of 'gay', prerequisite to the notion of its genetic inheritance, can nevertheless be disaggregated from the gene(tics). Jones resolves this tension in a claim for a homosexuality science that has no moral meaning; Connor and Wilkie argue that the science has, in and of itself, no moral implications.

Significantly, the neutralisation of the moral *status* of scientific knowledge depends here on a further disaggregation – a distinction made between scientific from moral *agency*. The 'we' of Connor and Wilkie's claim articulates a displacement of moral agency on to those ('us') who, specifically, do not exercise scientific agency. Since scientific *problem solving* is disassociated from *problematisation*, scientists cannot be held accountable either for the social consequences of their practices or for their own take-up of commonsense problematisations. Moreover, even as moral agency is dislocated from scientific agency, so too is the *moment* for moral considerations. As Jones suggests in his *Daily Mail* article 'Scientists cannot evade their responsibility; they, after all, have provided the opportunity for [moral] choices, but they must not be the people who decide. Society must do that' (Jones 17 July 1993). There is a rather stunning sleight of hand here, from an assertion of scientific responsibility, followed by its immediate denial – projected instead, *post-facto* on to 'society'.[23] Jones' perspective both here and in his statement in the *Independent* (quoted above) is significant for the way it seems, on the one hand, to locate science within a social context, while on the other, to posit that its *use* (or misuse) is an extra-scientific phenomenon. Implicit here is the rather odd suggestion that scientists, indeed scientists *alone*, do not 'use' the artefacts of their own practice.

The idealisation of this displaced 'post-facto' investment of scientific moral agency in 'society' emerged in a number of articles as a call for a (legislatively drafted) 'gene charter' to control the 'misuse' of genetic information. In the *Daily Mirror* for example, this call for a 'gene charter' was cited as a bulwark against the potentially monstrous implications of genetic science:

> Patrick Dixon, author of *The Genetic Revolution* says a "gene charter" was urgently needed to avoid abuses.

> "Every new gene discovery brings closer the horror of eugenics, designer families with embryos selected for intelligence, hair colour or sexual orientation" says Dixon. (Swain, 17 July 1993)

The 'gene charter', invoked here as the mechanism for the achievement of scientific balance, also bespeaks journalistic balance. The rational, democratic, public interest connotations attending the notion of a 'charter' not only

---

23 Connor and Whitfield (*Independent*, 17 July 1993), both validate and problematise this expectation of 'the public' to handle scientific knowledge by raising three issues: a) whether 'piecemeal' announcement of the results of genetic science fuels public resistance to confusion with respect to the science; b) the need for there to be a debate on 'what *we* regard as "normal" genetic information otherwise scientific results might be used to further polarise opposing factions still further' (emphasis mine); and c), quoting Paul Nurse, Professor of Biology at Oxford University, that delay in announcing scientific results was illegitimate: 'It won't help if there is a restriction [on science], he said'.

discursively neutralise the horror *narrative* but, in so doing, offer a symbolic containment of the material possibilities of an horrific science. It is significant that neither the *Mirror* nor any of the other stories taking up the 'gene charter' theme saw the need to provide substantive elaboration on what such a 'charter' would constitute or how it would work. The invocation alone seemed to be taken as enough to signal the containment of a safe genetics.

The final foreclosure on the attribution of moral meaning to scientific agency emerged in the reassertion of a characteristic trope of both utopian and dystopian narratives of scientific progress – that good or bad, it is inevitable:

> The question then is where and how to draw the line. Some would argue that scientists should stop meddling with our genes, that their research should cease immediately. *This, even if desirable, would be almost impossible to enforce.* But critics may well ask why 26 nations are contributing a total of $2 billion to attempt to map the entire human genetic make-up when their governments have not, apparently, given more than a second's thought to possible safeguards against *the results being misused.* (*Independent*, 18 July 1993, my emphasis).

The notion that even the undesirable activities of science are ultimately outside social control significantly undercuts calls for its social regulation. The pessimism of this passage is notable for the way in which it offers a critical recognition of science *as* a social–moral practice only to characterise that recognition as futile.[24]

Abortive Anxieties – Mothers Make/Break Gay Sons

As many of the passages quoted above suggest, it was specifically the fear of eugenic applications of the 'gay gene' science that constituted its perceived monstrous possibilities. Indeed, implicit in the understanding of a 'gay gene' as a diagnostic 'fact' is the suggestion of a 'therapeutic' abortion. Here the seemingly easy slippage between 'gay genes' and those for 'other diseases' could not but evoke both the material and conceptual terrain of prenatal diagnostic technologies. Juxtaposed with a scientific monster narrative (only partially contained by the displacement of moral agency on to 'society'), it was perhaps inevitable that abortion, with its own distinctive monster mythologies, would become a central discursive resource for the articulation of the 'gay gene' moral dilemma. In this context the spectre, as many papers had it, of 'gay

---

24 A 'progress is inevitable' assertion very often begins with (or comes with a corollary) elision of critical debate with a 'call for a ban'. Both are rhetorical formulations that tend to effectively (if not intentionally) foreclose on further discussion.

genocide'[25] became a point of convergence for a network of anxieties around diagnostic imperialism, women's reproductive choice and, saliently, the power of mothers to make and break sons.

Mothers Make Gay Sons – Recessive Narratives: and the Freud Connection

> The researchers concentrated on the X chromosome which men inherit only from their mothers, after studying family histories that tended to show that homosexuality was passed down through the maternal side. (Hawkes, *The Times*, 16 July 1993)

> Until these findings began to surface, it was generally assumed that homosexuality was a matter of free choice conditioned by upbringing. Parents may have been held to play a large role in this shaping of their children's sexual orientation. Freud and his followers have pointed accusing fingers at over-loving mothers and inadequate fathers. But parents were not thought responsible for actually passing on a tendency to homosexuality, like one to left-handedness. It is unfortunate that once again mothers seem likely to take the main blame. (*Independent*, 16 July 1993)

The preoccupation with pathological and pathogenic mothers in the face of 'deviations' from conservative familial discourse is as endemic a feature of popular representation as it is a cliché of psychoanalysis. The take up of this discourse in the search for 'gay genes' is precisely suggested in Hawkes' note of the scientific concentration on families where a maternal line of inheritance could be inferred.[26] Simon LeVay, whose earlier research on 'gay brains' set the stage for Hamer et al.'s study, was also quoted as being influenced by the 'Freudian view of the origin of homosexuality' (Connor, 16 July 1993, *Independent*). The notion that homosexuality is a recessive, maternally-inherited predisposition or trait both reinvests in and reifies the psychoanalytic narrative of homogenerative mothers. Indeed, 'gay genes' would seem to confirm the impossibility of true Oedipal separation. [27] Here the 'inadequate father'

---

25 For example, a passage from the *Independent* read 'Peter Tatchell, spokesman for the gay rights group Outrage, said that aborting foetuses that carry a genetic predisposition to be gay "is tantamount to prenatal genocide of homosexuals"'(Connor and Whitfield, *Independent*, 17 July 1993).

26 Ewing (1995) specifically critiques Hamer et al. for discarding subjects from the study who did not fit the pattern suggesting maternal inheritance.

27 In Lacanian terms, gay genes would seem to posit a son whose claim on the phallus is compromised, who can never fully enter the realm of the symbolic, tied as he must be to the Imaginary through unbreakable biological connection to the body of the mother.

becomes redundant as maternal influence appears to be singularly 'to blame' for the compromised masculinities of gay sons – mama's boys truly, after all. 'Gay genes', maternally inherited, also recapitulate the embodied effeminacies – limp wristed, feminine voiced – of popular stereotypes of gay men – intrinsically flawed through their direct corporeal links with femaleness.

Mothers Break Gay Sons – Endangered Embryos and Gay Genocide

> The research published in the learned American journal *Science* raises disturbing ethical issues. Could, for example, this discovery eventually lead to ante-natal tests being offered to pregnant women who might choose to abort a foetus carrying such a gene? The very idea fills me – and I suspect most people – with revulsion. (Jones, *Daily Mail*, 17 July 1993)

> Now Mums-to-be may be offered tests to discover if their boy is going to be homosexual and doctors fear parents could decide on abortion rather than have a gay son. (Watson, *Sun*, 17 July 1993)

> A programme of screening foetuses for the relevant gene and then offering mothers the possibility of abortion (as is done with major congenital defects like Down's Syndrome) would come up against rival claims of civil liberties. A woman's right to abortion on her own chosen grounds would have to be weighed against the potential right of a foetus to develop in its own way and by implication the right of other adults to live in the way they prefer without stigma. (*The Times*, 17 July 1993)

> A House of Commons Early Day Motion was put down yesterday by pro-life Liberal Democrat David Alton, and signed by 30 MPs, calling for a gene charter to tackle the problem.
>
> A Spokesman for the Campaign for Homosexual Equality warned; "Once you start offering the facility to choose babies of a certain hair colour, skin colour, sexual orientation etc, you are on a very slippery slope. Gay people have a much right to be born as anyone else". (Swain, *Daily Mirror*, 17 July 1993)

Monstrous mothers are stock characters in both master narratives of homosexuality and anti-abortion discourse. In both contexts, endangerment is constituted as intrinsic to the maternal body and will. As these passages illustrate, the take-up of anti-abortion repertoires in the 'gay gene' stories signals a number of startling narrative shifts. For example, mothers, in this context, become responsible not only for homosexuality, but for homophobia. The investment of 'society' in eugenic selections and the incipient homophobia of the 'gay gene' science itself are thus displaced on to women who given the choice would choose, it is asserted, to annihilate gays. This construction of homo-destructive mothers is entirely congruent with, indeed foregrounded through, the anti-abortion commonsense of pregnancy as an adversarial condition – 'two people in one body, one of whom is about to be murdered'.[28]

There is, in this formulation, a latent problematisation of the 'gay gene' science for its provision of what is perceived to be yet another (illegitimate) rationale for abortion.[29] As a consequence, the 'fact' of gay genes is implicitly recuperated. However, it is a woman's right to choose, rather than homosexuality science (or institutional homophobia), that is constituted as the harbinger of gay genocide, as the threat to gay rights 'to be born', 'to live the way they prefer, without stigma'. Hence the call for a 'gene charter' which pits gay freedom against women's choice and reconstitutes gay rights struggle as a corollary of anti-abortion politics. Gays, here, elide with endangered embryos, claimed as points of projective identification against the genocidal potentialities of diagnostic genetics. Indeed, in the face of such abortive anxieties, gays, even for the iconographically right-wing, gay-bashing *Sun*, become momentarily transformed from repressed 'other' into 'the-rest-of-us'.

Here There be (No) Monsters – (Re)Defining Homosexuality

Amidst the narrative oscillations between 'good' and 'bad' science and the displaced anxieties surrounding eugenics-minded mothers, the 'gay gene' stories appear to destabilise the characteristic disease-association of dominant discourses of homosexuality. Steve Jones, for example, stated baldly: 'Let us dispose of one issue straight away: the fact that homosexuality can be coded for by genes does not make it into a "disease" or a defect' (Jones, 17 July 1993, *Daily Mail*). Two articles from the *Independent* rather amusingly posited that gays ('gay genes') might even carry an evolutionary advantage:

> One suggestion [as to why the "gay gene" did not die out through natural selection] is that a genetic predisposition would somehow confer an advantage to sisters of the man, because in primitive societies he would be less likely to set up his own home and more likely to help raise his sister's children. But Professor Jones said there is "absolutely no evidence of this". (Connor, *Independent*, 17 July 1993)

---

28 This characterisation of pregnancy was asserted by Bernard Braine and other parliamentary supporters of the 1988 Alton Bill which attempted to lower the upper-time limit for legal abortion to 18 weeks gestation. For further discussion, see Steinberg (1991).

29 Within anti-abortion discourse, all rationales for abortion are illegitimate.

Some genes do give rise to what is inherently a disease, such as sickle cell anaemia. It might seem tempting to eliminate this gene, but it turns out [it] … also protects against malaria. What other functions might a gene for homosexuality perform? We have no idea. (Connor and Wilkie, *Independent*, 18 July 1993)

The 'great gays' argument, so characteristically prominent in the wake of Section 28 debates of the late 1980s,[30] was also reinvoked in resistance to the prospect of gay screening:

Leaving aside the Hitlerian eugenic attitudes implied, only a glance is required at the list of homosexual geniuses down the ages – from Michelangelo to Britten, Bacon and Nureyev – to show the catastrophic loss that [screening] would inflict on mankind and not just by eliminating geniuses. The contribution of homosexuals to making society more civilised and less brutal can hardly be overestimated. (*Independent*, 16 July 1993)

Yet as both the 'positive evolution' and 'great gays' positions illustrate, the destabilisation of the homosexuality-as-disease narrative was highly ambivalent. Not only are disease referents explicitly retained as analogies, as for example with the reference to cystic fibrosis, but homosexuality as a problem or abnormality, albeit benign, is reasserted. The evolutionary picture of 'successful' masculinity measured in terms of heterosexual dominance and reproduction is only rather dubiously liberalised with the possibility of gay men as helpful 'sisters' to their heterosexual sisters. The projection of the modern notion of homosexual identity into 'prehistory' has an interesting double movement suggesting homosexuality as, on the one hand, original to the human condition, and on the other, as an intrinsically and transhistorically subordinate form of humanity. In this context, even 'natural selection' takes on an ambivalent meaning, with the implication being that if nature did not eradicate this 'abnormality' (even if 'we have no idea' why), then neither should we.

Minor Monsters – From Disease to 'Minor Ailment'

Within the frame of liberal resistance to the eugenic screening of gays, notions of homosexuality-as-disease were largely retained, but in reconstituted forms and demoted in their importance. This was played out in a number of ways.

---

30 Section 28 of the Local Government Act 1988 is a vaguely worded proscription against Local Authoritiy's support of the 'promotion of homosexuality' or the validation of gay families (termed in the Act as 'pretended family relationships'). See Stacey (1991) for further discussion of the 'great gays' position taken up amongst activists and in the popular press in the wake of the Stop the Clause Campaign.

For example, even Steve Jones, in the very same article as his blunt rejection of the association, went on to make an analogy between the genetic predisposition to homosexuality with conditions such as diabetes, cancers, haemophilia and with the minor and correctable condition of myopia. Of the latter, which he himself suffers, he stated: 'Now I am old and know that [myopia] is genetic. It makes no difference to the treatment. I just carry on wearing the glasses' (ibid.). Indeed, the lateral associations in the flow of most articles placed homosexuality squarely in the frame of serious disease, even where there were graphic rejections of that formulation. There is, moreover, an interesting series of slippages in Jones' reference first to a 'gay gene' (where gay is not a disease), then to a slight abnormality, then to the notion of (small, adjustive) treatment. Indeed, this extraordinary reconstruction of homosexuality as a 'minor ailment' was a consistent theme across the 'gay gene' reportage:

> A British expert on medical ethics warns that should the theory prove correct and a prenatal technique become available to test for the gene, new laws may be needed to prevent the abortion of foetuses carrying it. "We are testing foetuses in many ways in the womb and are aborting them if they have sometimes *relatively minor abnormalities*" says Richard Nicholson, editor of the Bulletin of Medical Ethics … We must have a wider public debate about how much difference we will accept in society. (Swain, *Daily Mirror*, 17 July 1993, my emphasis)

> Richard Nicholson, editor of the *Bulletin of Medical Ethics* said he hoped the debate would highlight the fine line between *minor abnormalities in the foetus that should not lead to termination* and more serious problems that did. "I don't regard homosexuality as an abnormality that needs to be changed back to heterosexuality". (Connor and Whitfield, *Independent*, 17 July 1993, my emphasis)

> It is here that we begin to understand that this is not an issue for homosexuals alone. One in 30 children is born with a genetic problem of some kind – blindness, deafness, mental handicap for example. Perhaps we can accept the rights of their parents to decide that these are handicaps with which they would rather not cope. But what of a gene that predisposes to cancer or heart disease? … And that leaves aside all the other things that cause children to disappoint their parents. (*Independent*, 18 July 1993)

The emphasis in these passages on tolerance of (or, at the greatest extreme, treatment for) homosexuality is clearly underpinned by an understanding of abortion as the greater social ill. It is also possible that a more general liberal shift in popular sensibilities around homosexuality might have foregrounded the possibility that screening out homosexuality could be perceived as 'not an issue for homosexuals alone'. Whatever the case, the relegation of homosexuality to 'minor disease' was widely asserted, with the effect, at the very least, that *eradication* could not be regarded as an appropriate response to homosexuality. It is, perhaps, not surprising that treatment (including gene therapy) emerged, in the tabloids to resolve tensions between competing monster narratives of eugenic science, homogenerative/destructive mothers and abortion:

> We are not prisoners of our genes. Science can reveal what is in them, but it can also treat and cure what it finds – if society decides that it should. Genetic research is only the beginning of that process, not the end. (Jones, *Daily Mail*, 17 July 1993)

> Recent advances have made gene therapy a reality. The treatment has the potential to help thousands of sufferers of genetic diseases caused by single gene defects. Doctors hope one day to treat cancer by manipulating genes. (Watson, *Sun*, 17 July 1993)

> Dr Tony Vickers, former head of Britain's work on the Human Genome Project, says the ethical issues involved in gene therapy are not only which problems should be tackled but also ensuring that any successful therapies become widely available. He says most countries resist any genetic engineering that would result in permanent and inheritable changes to genes, as opposed to therapeutic work. (Swain, *Daily Mirror*, 17 July 1993)

The 'treatment' option, though only laterally associated with but not explicitly linked to 'gay genes', clearly reinvests in the master narrative of homosexuality while disclaiming its most extreme implications. Indeed, against the possibility of 'gay genocide', the invocation of treatment represents the possibility of (if not the call for) a more tolerant view of homosexuality as well as a more balanced, even anti-homophobic science. In the broadsheets, by contrast, 'treatment' was more likely to be characterised in similar terms to eradication. Connor, writing for the *Independent* for example stated: '[a] principle concern is that people who perceive homosexuality as an abnormality will call for attempts to diagnose it, perhaps by a pre-natal test, or attempt to "cure" it by developing so-called treatments' (Connor, 16 July 1993).

If the elision of anti-eugenic–anti-abortion discourse framed a highly contradictory articulation of resistance to homophobic selection (and to the disease-construction of homosexuality), so too, at moments, did forms (or disclaimers) of racism.

> Does this mean being gay is a disease?

No, this discovery merely means that the variation in human behaviour mirrors to some degree the variation in human DNA ... Some variation is completely determined by the genes; whether someone has blue eyes or brown. But no one suggests having brown eyes is a disease or that being blue-eyed has greater moral value. (Connor and Wilkie, *Independent*, 18 July 1993)

Different societies have always held different beliefs about some or all of these conditions. In India, thousands of abortions are carried out each year for the simple genetic reason that the foetus is a girl. To me, this is absolutely wrong, but I say this as a human being, not as a scientist, and my views represent those of the society I come from which, thank god, had advanced way beyond that. In modern Britain, homosexuality is a legal, acceptable preference. Science neither can, nor should, alter that. (Jones, *Daily Mail*, 17 July 1993)

The disingenuity of the references to and rejection of racial hygiene ideology implicit in both these passages is striking. While it may be true that blue eyes, for example, have not been associated with disease, brown-eyed people have frequently been subject to pathologised discourse and blue eyes have most certainly been invested with notions of both moral and racial superiority. Connor and Wilkie's ill-chosen analogy (and wishful thinking) here seems to reinforce rather than disclaim the very embeddedness of moral discourses that accrue even to seemingly innocuous physical characteristics. Moreover, the attempt to redefine homosexuality as a morally neutral behavioural variation is significantly compromised by its reinvestment in biological discourse. Perhaps more disturbingly, Jones' explicit anti-homophobia, claimed as a modern British commonsense, is articulated through a graphically colonialist (Orientalist) evolutionary discourse. In this formulation, science is characteristically re-located within civilised sensibilities; monstrous 'othering' is, reassuringly, produced only by (other) monstrous 'others'.

The Discrimination Debates

The partial destabilisation of the disease discourse of homosexuality was perhaps most emphatically illustrated by the preoccupation in all the papers with the question of homophobic discrimination and gay rights. Aside from (although also incorporating) the question of eugenic screening, significant attention was given to debates about whether the gay gene discovery would intensify the social stigmatisation of homosexuality or, alternatively, would provide a 'natural' basis for the extension of civil rights protections.

The happiest outcome of the latest research would be to reduce the intolerance that homosexuals encounter in virtually all fields except show business and the rag trade. Religious fundamentalists would find it harder to label homosexual behaviour a sin. Parents could cast aside fears of their children being corrupted by gay teachers. More importantly, homophobes might come to appreciate that sexuality is an infinitely complicated matter, in which nature and nurture play roles that are likely to remain hard to unravel. (*Independent*, 16 July 1993)

Michael's reaction to news of the gay gene was double-edged. On the one hand he fears there could be some lunatics who would take this discovery as an excuse for arguing foetuses carrying the gene should be aborted. On the other, it should make no difference since homosexuality ought to be accepted as natural anyway. But if the scientists prove their theory, it could make a profound difference to family relationships. "It might encourage young people to tell their parents. They should know they haven't done anything wrong". (Swain, *Daily Mirror*, 17 July 1993)

Even among gay activists there is disagreement about whether proof of a biological tendency would result in more, or less, discrimination. On the one hand, the fact of genetic basis would undermine the idea that homosexuality is a form of moral turpitude … In these terms, demonstrating that a homosexual life is not chosen in free will would result in more protection and less vilification. But those who defend the idea of homosexuality as an alternative lifestyle … reject such biological determinism. They do not wish to be excused their sexual orientation on the grounds that they cannot help it, but to be accepted as they are. There is a fear that those who regard homosexuality as inherently deviant and unacceptable could use such a scientific breakthrough as a way of eliminating the tendency altogether. (*The Times*, 17 July 1993)

Whatever the ambivalence surrounding the causes and status of homosexuality (and in part made possible by it), the discrimination debates emphatically constituted *homophobia* as a problem – which the finding of 'gay genes' would either ameliorate or add to. In all three passages, it is the homophobe, rather than the homosexual, that emerges as disclaimed 'other' – 'lunatic' at worst, irrational at best – who embodies the threat of gay discrimination. Indeed, the homosexual monster narrative itself, with its characteristic themes of 'sin', 'corruption by gay teachers', doing 'wrong' to one's parents, is graphically reconstituted as a *homophobic* monster narrative. In this context, gay rights activists (or their supporters) are discursively repositioned, both in terms of victimhood *and* affirmative social agency, onto the moral high-ground. Moreover, although displaced on to the putatively non-scientific figure of the homophobe, homophobia nevertheless constituted a more or less implicit yardstick against which the 'gay gene' science was, in the immediate sense, to be judged, and a symbolic index of judgement about the multi-discriminatory potentials of genetics more generally. Connor and Wilkie take up this point directly:

> Gay activists in Britain however, are suspicious of any scientists who make a study of homosexuality, they fear they are being singled out and labelled as a problem because of perceived abnormality. They fear that homophobic elements in society will use the genetic research to "diagnose" or "treat" homosexuality as if it were a disorder to be cured or eradicated. (Connor and Wilkie, *Independent*, 18 July 1993)

At the same time, in their consideration of the potentially beneficent, anti-discriminatory effects of the 'gay gene' discovery, these debates effectively recuperated the science. For example, for the purpose of both sides of the debate (i.e. the claims of a natural basis for the extension of rights or of a dangerous knowledge), 'gay genes' had to be taken as 'fact'. The emphasis, moreover, on the (mis)use of the 'gay gene' science by 'homophobic elements in society' also reinvested in the notion of a disinterested science even where the (justified) suspicions by gay activists of scientific motives in the study of homosexuality were noted. Finally, the ascendancy of anti-abortion discourse with its attendant stigmatisation of monstrous mothers, was implicitly reconfirmed. Whether 'gay genes' were seen to potentially fuel a 'final solution' to the 'gay question' or to reinvigorate gay rights struggle, a woman's right to choose remained on the wrong side of the debate.[31]

Discursive Displacements

Lisa Smyth has traced evolving social theory on the phenomenon of the *moral panic*. As she has noted in her study of *the x case* – where x was a 14-year-old rape victim, prevented from travelling to have a pregnancy termination – the notion of a 'moral panic' is generally taken to signify 'the mass media's involvement in

---

31 Connor and Wilkie do suggest another danger in this context: 'But what would happen to the child if its parents decided not to abort after a "positive" test? It might grow up knowing that it had been labelled as genetically gay, a burden of knowledge that would inescapably affect his future life and choice of sexual partner. It could become a self-fulfilling prophecy'(18 July 1993, *Independent*). Interestingly, this passage recuperates the homosexual-as-problem narrative (as well as the science-as-problem narrative), by introducing a twist on the anti-abortion theme – the danger of prenatal diagnosis when abortion is refused. It is significant here that commonsense assumptions of heterosexuality are not perceived as a 'burden of knowledge', though such assumptions as we all know are often (wishfully) regarded as 'self-fulfilling prophecies'!

maintaining and policing definitions of [certain] social behaviours [or groups] as "deviant" or problematic' (Smyth 1999, p. 1). The role of moral panics in shoring up and recuperating the hegemonic grip of social conservatism on popular commonsenses of 'law and order' and 'family values' has been widely examined.[32] Indeed, it can be argued that they constitute, in this capacity, key *pedagogic* strategies of the social-conservative Right. Moreover, as Epstein and Johnson (1998) have argued, moral panics are perhaps best understood as *defensive* responses to perceived or actual erosions of the fragile hegemonies of conservative philosophy, or, to borrow from Raymond Williams (1977), moments where the recuperation of the dominant is staged on the terrain of the apparently emergent.

Yet the reportage surrounding both *the x case* and the 'discovery' of a 'gay gene' depart from the characteristic narrative sequences of a moral panic, even as their meanings and forms seem to be animated from within its popularly recognisable conventions. In *the x case*, as Smyth notes, there was a seemingly reverse trajectory, a moment where the tensions of social conservatism seemed to become unbearable, where the dominant commonsense broke down and the emergent resistances of the liberal left appeared not only to defeat the right wing consensus on both abortion and Irishness, but indeed generated a significant (though only partial) reversal of constitutional law. Likewise, the press coverage of the 'gay gene' in Britain also represents a moment of emergent liberal rather than dominant conservative sensibilities. Here the familiar moral panic about homosexuality gave way to the *moral dilemma* where competing narratives of homosexuality, science and abortion became enmeshed in what might be termed a *circuit of displacements*.[33] The continuous slippages between the problematisation of homosexuality on the one hand and of homophobia on the other were embedded in further narrative oscillations pitting gay rights, for example, against abortion rights, monstrous science against monstrous mothers, scientific against moral agency. Yet even as homophobia emerged as a key object of critical focus, conventional constructions of homosexuality were retained. The notion that homosexuality is abnormal, for example, was reasserted in its representation as a static biological–social category, a statistical minority and, particularly, a recessive trait. While the view of homosexuality as a natural disaster was refused, it remained, at best, a socio-biological idiosyncracy

---

32 See, for example, Cohen (1972); Hall et al. (1978); Epstein (1993; 1997).

33 I am borrowing here from Richard Johnson's concept of a 'circuit of production' which refers to the continuous interplay of productive-interpretive moments (authorship and readership) in the making of meanings within particular and competing discursive fields (Johnson 1986). The 'circuit of displacements' is intended to describe the continuous discursive slippages produced through the specific narrative convergences in the 'gay gene' stories.

that at the least required explanation and possibly also required tolerance both from 'society' and science. The presumption of heterosexuality as normative, normal (and even an inherited characteristic) went unquestioned.

The *liberal-resistant* reframing of homosexuality/homophobia was grounded in both the ambivalent status of gene science (including its associations as monster science) and the power of anti-abortion commonsense. In this context, the reconstitution of gays as points of projective identification relied on the displacement of monstrous anxieties (about gays, genes and eugenics) onto women. Thus, contained within (and productive of) the moral *dilemma* about 'gay genes', was a residual moral *panic* about monstrous mothers. Ultimately, the gay gene stories represent not so much an encouraging shift in popular sensibilities around homosexuality (though they do suggest this possibility), but rather that in clash of monster stories, some monsters might become, if not heroes, then at least not quite so monstrous.

Chapter 5
Trace: On Genes and Crime

Introduction

Crime, justice and jurisprudence are evocative and particularly salient fixtures, as much of the cultural imaginary as they are as of social praxis. Scientific discourse, moreover, has been central to the development of modern criminology, constituting its theoretical underpinnings and informing social and governmental policy, as well as the genre conventions of criminologically-themed representation. As a point of focus, the articulation of genes and crime draws together a number of vexed and highly complicated scientific and social questions. These include, on the one hand, the putative biological dimensions of behaviour, context and identity and, on the other, socio-legal questions concerning the meaning of justice, the categorisation of 'crime' and the jurisprudential legitimacy of evidence and evidentiary extrapolation. And, of course, as with the other subjects explored in this book, crime constitutes a multivalent *signification field,* and a particularly important one, not only for the cultural currency of genes and genetics, but also for the terms of their persuasion and plausibility.

This chapter[1] will explore the inter-resonances of three case studies in which genes and crime conscript (and are conscripted into) the wider cultural imaginary. First is *The Genetics of Criminal and Anti-Social Behaviour,* the 1996 published proceedings of an early and foundational scientific conference and part of the CIBA Foundation series. The papers in this collection are concerned with the capacity of genetics to both explain crime and to carve out a mode of predictive utility in the social field. Second is *The Innocence Project,* a legally-framed social action group and undertaking whose mission is to exonerate those who have been illegitimately imprisoned for crimes they did not commit. The *Project* has been particularly galvanised by the advent of DNA fingerprinting and the potential of genetics to remedy the multivalent inequities of the criminal justice system. Third is the internationally syndicated television franchise *CSI,*

---

1 Some arguments in this chapter appeared in earlier formulations, in Deborah Lynn Steinberg (1997) 'Genetics of Criminal and Anti-Social Behaviour', *Sociology* 31(1): 187–9.

a forensically focused drama in which genes and genetic fingerprinting figure centrally and which has played a powerful role in sedimenting genetics into everyday vernacular and into the wider cultural imaginary. As I shall argue in what follows, all three of these case studies represent what might be termed *epistemic genres*, that is, contexts and discursive representations explicitly concerned with the nature of knowledge and, specifically, with the relationship between truth and evidence. Furthermore, all three contexts are concerned with *justice* as a medico-moral (or perhaps more accurately *epistemo-moral*) enterprise and have directly informed the connotative currency of genes as what might be termed *ethical facts*.[2] Finally, all three case studies are concerned with the relationship of science to governance and governmentality, albeit, as we shall see, with divergent levels of reflexive interest in science itself as a site of governmentality.

This chapter, then, extends focus on the conditions of *possibility* of knowledge – the power relations of its context, its socio-cultural consequences and its modes of iteration (material and representational) – to an additional question; that of the conditions of *plausibility* of knowledge. If the first part of the book concerned regimes of truth, the second part will consider as well, its terms of persuasion. As I will suggest over the next few chapters, what constitutes the genetic *episteme* is constituted as much by the *affective* – that is, the realm of *feeling* – as it is of what has conventionally been understood as epistemic.

Genre and Episteme

In the study of media, *genre* characteristically refers to conventions of style that are seen to define particular classifications or types of media fare. These include considerations of narrative arc, visual–aural aesthetics and particular signification codes and repertoires.[3] Film and television, for example, are typically classified in broad categorisations referring to both format and conventions of content – drama, comedy, western, thriller, science fiction, police procedural, rom-com and so on. Genre can refer to sub-generic characteristics – splatter films (a variant of the horror genre), or bromance. Genre considerations can also refer to modes of address, production values

---

2 Here I am extrapolating from *The End of Faith*, by Sam Harris (2006). Harris' book is concerned with religion and faith-based (as distinct from empirically-based, scientific) epistemology. His discussion of knowledge, good and evil suggests a distinction between what we might term a 'utilitarian fact' and an 'ethical fact'. This is a way of understanding truth as situated, and specifically, that truths emerge from and fuel particular kinds of power.

3 For extended discussion of genre and genre studies, see, for example, Neale (1990); Long and Wall (2009); Braudy and Cohen (2009).

or context and social capital, for example, distinctions between 'academic' and 'popular'; 'news' and 'propaganda'; documentary and 'constructed reality'. As an arena of theory, genre studies has been strongly inflected by theories of 'the gaze'[4] and considerations of specular pleasure. In all of these contexts, *genre* describes what might be usefully understood as a contract of intelligibility between producers and consumers – that is, a normative set of expectations: whether these concern sequences of plot; modes of visual spectacle; or values as 'fact' or 'fantasy' – that shape the 'grain' of meaning. It is also important to note that the boundaries of genre categorisation on any of these counts are often blurred; many media forms defy easy categorisation, often drawing on elements of multiple genres.

The case studies I examine in what follows exemplify all of these definitions, crossing over between scholarly-scientific output, socio-political project (with social activist modes of discourse and communication) and procedural-forensic entertainment. What binds them are a set of shared *interpretive-imperative* fields – crime, justice and genetics – and explicit concerns in each context with science, ethics and the plausibility of knowledge. They are all, notwithstanding their generic distinctions in other terms, modes of *epistemic genre*. To examine them from this vantage point foregrounds a number of questions that perhaps less typically arise in studies of signification: those concerning epistemic premises (the foundational assumptions of knowledge claiming) and extrapolations (modes of deduction and logical inference) and the socio-political praxes of signification (the 'languages' of meaning-making that directly mediate – and are mediated by – the formal exercise of power).

Science of Crime

Genes are likely to influence the occurrence of criminal behaviour in a probabilistic manner by contributing to individual dispositions that make a given individual more or less likely to behave in a criminal manner. (Lyons 1996, p. 61, CIBA Foundation Symposium 194)

[I]t is unlikely that there is a single gene for crime *per se*, but genetic factors are likely to play an important role in behaviours, such as aggression or impulsivity, that are associated with or predispose to crime. (Silberg et al. 1996, p. 77, CIBA Foundation Symposium 194)

I was interested in the relationship between maternal psychopathology and child crime. If the mothers of the criminals suffer from an elevated rate of antisocial

4 See Mulvey (1988).

personality, you might have attributed the relationship to genetics. (Mendick, in discussion, p. 88, CIBA Foundation Symposium 194)

We are not looking for heritability here, but for specific genes related to aggression. (Carlier 1996, in discussion, p. 96, CIBA Foundation Symposium 194)

This chapter concludes by emphasising that the question is not whether genetic evidence will ever be admitted into court, but when and under what kinds of circumstances. No doubt, genetic evidence, and comparable kinds of biological evidence, will have a major impact on juries when such evidence is more fully accepted by the legal and scientific communities. (Denno 1996, p. 248, CIBA Foundation Symposium 194)

Genetics of Criminal and Antisocial Behaviour, published in 1996, is a collection of papers, including transcripts of discussion, from a symposium of those interested in behavioural genetics as applied to crime. It thus represents an early and agenda-setting conference for an emergent field specialisation. The symposium was sponsored by the London based CIBA Foundation, an independent philanthropic offshoot of the CIBA-Geigy Pharmaceutical company. The CIBA Foundation's stated mission is 'to promote international cooperation in biological, medical and chemical research';[5] their symposia characteristically explore moral–social questions as part of scientific discussion.

As a scientific discussion forum and publication, *Genetics of Criminal and Antisocial Behaviour* is part of a larger spectrum of scientific exchange, spanning academic journal publications at one end and (arguably) science journalism at the other. What distinguishes the conference context, and even more strongly in the case of journal publications, is peer-review: a system of certification aimed to safeguard the integrity of scientific knowledge production.[6] Scientific fora thus constitute an *epistemic genre* both in terms of their foundational project

---

5 The fly-leaf of their publications describes the organisation and its mission as follows: 'The Ciba Foundation is an international scientific and educational charity. It was established in 1947 by the Swiss chemical and pharmaceutical company of CIBA Limited – now CIBA-GEIGY Limited. The Foundation operates independently in London under English trust law. The Ciba Foundation exists to promote international cooperation in biological, medical and chemical research. It organises about eight international multidisciplinary symposia each year on topics that seem ready for discussion by a small group of research workers. The papers and discussions are published in the Ciba Foundation symposium series. The Foundation also holds many shorter meetings (not published), organised by the Foundation itself or by outside scientific organisations'.

6 Peer-review is itself a field of contestation. For extended discussion see Shatz (2004).

(to produce and communicate knowledge) and in terms of their professional and public capital. One of the historical conceits of peer-review, and of the scientificity it represents, is that politicisation undermines the scientific endeavour. The uneasy relationship of science and corporate sponsorship, for example, plays out precisely on the terrain of this crossover, as in the high profile examples of the sciences of food (sponsored by the fast food industry), tobacco (sponsored by the tobacco industry) and GMOs (sponsored by Monsanto). Yet science is part of a political ecology, even where it is uncontroversial. Genetics, as we have seen in earlier chapters, is a scientific context that has remained highly controversial even as it has pervasively sedimented and normalised. This is perhaps particularly the case where questions surrounding genes also include issues that are already profoundly troubled (race, class, gender and equality) or are already insecurely aligned with bodies, such as behaviour or identity. As with an earlier (1986) CIBA conference publication, *Human Embyro Research: Yes or No?*, controversy and a charged political context was an explicit terrain of *Genetics of Criminal and Antisocial Behaviour*. In the light of this, both collections, in so far as they argue their respective cases, can be understood both as *jurisdictional* (i.e. making rhetorical–territorial claims) and as *pre-emptive* (defending against a climate of public criticism or against a perceived risk of public backlash).[7] The other issue arising for scientific work, perhaps most acutely in the criminological context, is the problem of *confirmation/cognitive bias*. *Confirmation bias* refers to the tendency to find most plausible those knowledge claims that confirm one's assumptions. In the scientific context, it refers to the ways in which prior assumptions constitute either or both the constitution of inquiry and interpretive deduction applied to it.[8]

---

7 Public backlash is not simply a matter of public opinion (to the extent that this is reflected/fostered in popular media), but involves questions of law and institutional support. For example, human embryo research was embargoed in the US under the Bush administration; while in the UK it was regulated and funded.

8 A 2013 *Guardian* article (Jha, 26 August 2013) comments on joint research from the University of Edinburgh and Stanford University (Fanelli and Ioannidis 2013) that tracked the effects of pressure to produce 'exciting' results, which the researchers argued had particularly infiltrated US scientific research culture and the 'soft' sciences (including the biology of behaviour). As they point out, *confirmation bias* is not a description of conscious fraud, but of the ways in which the framing and context of research constitute an unconscious tendency of bias for the researcher. The article quotes University of Cardiff researcher Chambers, who suggests: 'My belief is that US scientists aren't actively engaging in mass fraud – instead, most of these questionable practices are unconscious … It's easy to fool ourselves into thinking that a result which "feels" right is in fact true. This problem is known as confirmation bias, which ironically was itself discovered by psychologists'.

The CIBA collection contains 14 papers, presented in clusters of three to four (presumably as they were presented at the conference), and interspersed with unmediated transcripts of general discussion. It covers a diverse range of topics spanning: the genetics of mouse aggression; the genetic significance of 'criminality' (equated most often with conviction records) and 'anti-social behaviour' among twins or adoptees; genetic 'predisposition' to crime; genetic mechanisms in 'impulsive behaviour', 'abnormal behaviour' and 'alcoholic violent' offenders; antisocial behaviour in evolutionary adaptation; the genetics of violence and warfare; and the legal implications of genetics and crime research. Contributors include genetics researchers, legal and bioethics scholars, and academics located in departments of psychology, psychiatry, and forensic psychiatry. The stated purpose of the symposium (and the book) are twofold: firstly, to present research which examines the 'contribution of genetic factors' to 'criminal and anti-social' behaviour (p. 1); and secondly, to elaborate a rationale which justifies genetic criminological research specifically and, by implication, behavioural genetics more generally. In both contexts, ethical, legal and socio-cultural implications of such research provide, with the marginal exceptions of Glover's discussion of genetic determinism, and Denno's examination of genetics and jurisprudence, are largely rhetorical.

The incipient positivism and essentialist tendencies of genetic science (discussed in various respects so far in this book) tap into wider debates about the (ir)reconcilability of biological frameworks with social issues, problems and processes. The putative genetics of 'criminal and anti-social' behaviour are a case in point. For example, across the range of papers presented at the CIBA Symposium, it is acknowledged that 'criminality' and 'anti-social' are cultural constructs whose meanings are historically both diverse and specific. Indeed, this point is set out at some length in the introduction to the published proceedings (Rutter 1996). At the same time, these terms are invoked throughout all of the presentations – often interchangeably – as if they described the same thing, and as if they had static, factual meanings from which common genetic factors can be deduced. As such, genes are both presumed *and* surmised as causal factors or embodied determinants for behavioural outcomes ('crime', violence, war, 'abnormal behaviour') which are in turn grouped together without explanation and implicitly reified into measurable singularities. Is all violence the same violence? Is all war the same war? Is 'anti-social behaviour' the same as 'crime' – what about where sociality is, itself, and what is criminalised – for example where gay relationships are illegal, in the history of miscegenation laws, or the criminalisation of girls' education or women's ability to travel independently? These ambiguities and elisions demonstrate what is perhaps the quintessential tautology of *confirmation bias* – that is, where a premise (genes link to/cause crime) is argued to prove itself.

Proctor (1988), moreover, has argued that the role of *apology* – that is the explanation of social processes as natural facts (with natural origins) – is a key characteristic of modern science which operates to naturalise social divisions and inequalities and police the status quo. In *Genetics of Criminal and Anti-Social Behaviour*, genes and their putative carriers are effectively construed both as 'criminal' and 'anti-social' bodies *per se* and as vectors for the transmission of social ills. Such a formulation, with its implicit elision of class in the epidemiological sense with class in the sociological sense, recycles scientific (eugenic and criminological) discourses of crime and poverty dating back to the nineteenth century. This subtext is particularly evident in the discussion sections which are punctuated with references to pathogenic mothers, the inner city and the socially deprived. Thus, for example, none of the studies refer to white-collar crimes, or to large-scale violent crimes committed by the affluent or extremely powerful. It is interesting in this context, that while many authors note that genetic research has a danger of contributing to negative social labelling and discrimination, none go on to consider how genetic research can be constructed either to avoid or to mitigate against such adverse consequences.

It must be noted that none of the authors posits a single gene for 'criminality'. Instead, all posit that a constellation of genetic factors – which on their own or in themselves are not necessarily productive of 'criminal' traits – interact with a complex of environmental factors to produce 'anti-social' or 'criminal' behaviour. As Glover (a symposium participant) argues, this represents a shift away from what he terms the 'hard determinism' of the single gene as cause paradigm, a paradigm which he and the rest of the contributors hold to be overly reductionist and scientifically untenable. All seem to support a 'soft determinism' which posits that notwithstanding their complex interactive character, biological 'risks' of 'criminal' and 'anti-social' behaviour can nevertheless be distinguished from social 'risks' (and, indeed, that in this way social 'causes' can be deduced from biological research). Glover states:

> I assume that the vast majority of scientists believe that genetic determinism is a rather extreme thesis and rather implausible for explaining the great range of human behaviour. But there is a much more plausible wider view which could also be called determinism, which is simply that genes together with other factors, such as neurodevelopmental and environmental ones, determine what we do. On this view, if we knew enough about genetics and neurobiology, and we knew the genetic make-up of an individual and about his or her environment, in principle a full causal explanation of behaviour would be possible. (pp. 237–8)

What emerges here is a more complex but not appreciably less mechanistic determinism. Glover defends 'soft determinism' because it, as he sees it, allows one to posit biological causality without denying moral agency; i.e. having 'criminal' predispositions does not negate criminal responsibility.[9] While the jurisprudential implications of genetic determinism (hard or soft) are clearly an important issue, Glover's critique side-steps the fundamental question of how a divergent cultural construct like 'crime' – which he, as the others do, invokes without defining it, as if it were a monadic behavioural (legal, penal and even philosophical) category – can be reconciled with notions of biological predispositions. Perhaps more critically, he assumes that it is possible to 'know enough' about these complexities such that a 'full causal explanation of behaviour would be possible'. The critical question of what threshold would define 'enough' to justify the deduction of cause (let alone a 'full causal explanation of behaviour') is left, again to tautology.

The collection vacillates significantly between claims of certainty and uncertainty. For example, there is debate about whether the value of genetic research in this arena is predictive or probabilistic with respect to propensities to commit crime in individuals; whether the question is of crime or of its (assumed) constituent affective-behavioural orientations (aggression, violence, impulsivity); whether genetics has the power to 'shed more light on causal processes [as opposed to] predictive strength in the population as a whole' (p. 11). As noted, most of the papers are prefaced with acknowledgement that 'crime' and 'antisocial behaviour' (if not 'aggression') are far from universal social categories. At the same time genetic inference is typically taken as proven. And indeed, certainty informs two underlying premises across the collection – one focused on the science and the other on its public interest. First is the premise that genetics plays a causal role in crime, notwithstanding the latter's slippery social interpretations and even if the mechanisms have not been defined (it is simply a matter of pinning them down). Second is the premise that genetic research *will* contribute to the social project of alleviating (or preventing) crime. The question is 'doing it right'. As Rutter (1996) sets out in the collection's introduction:

---

9 A Season 11 episode of *CSI*, 'Targets of Obsession' (Screened in the US on 17 February 2011) takes up this theme explicitly as it is revealed that CSI Langston and serial killer Haskell share the same MAO-A gene (a putative marker for a propensity to violent behaviour). The consequent proposition – that genetics is not social destiny – is offered both as dénouement and narrative twist. Specifically, the plot revolves around the question of biology, crime and criminal responsibility as argued through a court confrontation. The defense argument is that Haskell is not criminally liable because he has the MAO-A gene and therefore could not help what he became. The jury is not persuaded; the clinching point being the shock revelation that Langston has the same MAO-A gene. Significantly, and this is a point I shall return to later, it is in this narrative dramatic setting that the drive to certainty in genetics is, itself, a subject of tension. The narrative switch relies precisely on our understanding of the deterministic logic of genetic diagnosis – the genetic set up is (elided into) the generic set up. Such is the power of dramatic rupture; justice can lie in the uncertainty of science *because* of the uncertainty of biology.

The genetic study of antisocial behaviour should have practical benefits just because, if undertaken in the right way, the findings will aid our understanding of causal processes. That is crucial for the development of effective means of prevention or alleviation. (p. 8)

In addition to its underpinning conceptual contradictions, *Genetics of Criminal and Antisocial Behaviour* reveals important methodological problems. Many of the studies discussed in the book are based, for example, on demographic surveys that invoke crime statistics without explanation or interrogation of the indexical terms. For example, data comparing relative rates of 'criminal' behaviour (derived largely from conviction records) between men and women do not clarify what their 'crimes' actually were. Moreover, police records of arrest and conviction are taken at face value as indices of both criminality and criminal personality without consideration of the primary mediating role of policing, legal or judicial practice (let alone wider social inequalities) in who gets convicted and for what.[10] The chapters on the genetics of mouse aggression similarly avoid clarification of key questions. For example, what specifically does mouse aggression have to do with 'crime'? Indeed, what does human 'aggression' have to do with 'crime' and what do we mean by these terms? And given that the mice in question were conditioned to display aggression, how can genetic explanations of their behaviour be deduced or sustained? Interestingly, while gender difference was noted by most contributors as a consistent and dramatic point of distinction with respect to 'criminal' behaviour, none went so far as to theorise that maleness in itself constituted a predisposition to crime. Moreover, concomitant with unsustainable equivalences made between male and female 'crime' rates was a notable absence of consideration of normative male violence against women (which often does not make the crime statistics). Indeed, Rutter (1996) makes the extraordinary claim that 'stable marriage' mitigates against 'criminal' behaviour.[11]

Ultimately then, *Genetics of Criminal and Antisocial Behaviour* takes for granted precisely what it purports to explain, namely: how genetic explanations of variably defined social constructs like 'crime' and 'anti-social' behaviour can be

---

10 Lyons states: 'As mentioned [in my discussion above], genetic influences on intelligence might mediate the association between genetic factors and being arrested' (Lyons 1996, p. 67, CIBA Foundation Symposium 194).

11 Rutter states: 'Thus … a stable marriage and steady employment are protective against, and alcoholism a risk factor for, continuations in crime' (p. 6). This claim, which does not distinguish by gender, is also belied by patterns of domestic violence which do not suggest marriage functions to mitigate crime. It is also notable that domestic violence as a crime is subject to significantly different social and legal definition and status in different regions and nations, even as patterns of behaviour and the gender profile of perpetrators and victims remain comparable.

sustained; how even 'soft determinism' can be reconciled with the complexities of social behaviour; how genetics research can be shaped so as not to contribute to negative labelling and discrimination; and finally, what specifically the implications of such research are (or can possibly be) for, as invoked in the book's blurb, 'crime prevention and rehabilitation of offenders'.

Trace

> The Innocence Project is a national litigation and public policy organisation dedicated to exonerating wrongfully convicted individuals through DNA testing and reforming the criminal justice system to prevent future injustice. (*The Innocence Project*: home page <www.innocenceproject.org>)

If the CIBA Symposium focused on the genetic biology of 'crime' within a foundationally diagnostic framework, *The Innocence Project* draws on a different and arguably more plausible one: the evidentiary question of (culpable) presence at a crime, as established – or, more precisely, as falsified – through genetic means of identification. Founded in 1992 at the Benjamin N. Cardozo School of Law, at Yeshiva University, the *Project*'s work has underpinned the exoneration of over 300 convicted prisoners, with all but six based on post-incarceration submission of exculpatory DNA evidence. The *Project*'s website provides extensive details of their case files and its mission. A section of the website is dedicated to explaining the multiple ways in which convictions can be insecure (eyewitness misidentification, forensic error) or tainted (false confessions, investigative misconduct, incompetent or inadequate legal representation). This is countered, in a separate section entitled 'fix the system', with a detailed programme of proposed reform to the US criminal justice system that spans both legal and policy measures to establish and strengthen investigative, judicial and forensic oversight. This includes a call for the systematic use, across all jurisdictions, of DNA testing (which is not currently a standard part of criminal investigative due process). On DNA testing itself, the *Project* website states:

> Despite the widespread acceptance of DNA testing as a powerful and reliable form of forensic evidence that can conclusively reveal guilt or innocence, many prisoners do not have the legal means to secure testing on evidence in their case.[12]

---

12 The *Project* website sets out a remit for DNA testing legislation:
An effective post-conviction DNA access statute must:
  – Allow testing in cases where DNA testing can establish innocence – including cases where the inmate pled guilty.

The Innocence Project provides an interesting counterpoint to *Genetics of Criminal and Antisocial Behaviour.* First and perhaps most important is its particular relationship to the problem of confirmation bias. Where, for the CIBA symposium, the confirmation bias of the justice system is almost entirely ignored, even as its consequences provide the substantive underpinning of genetic investigation, *The Innocence Project* takes a critical counter trajectory. DNA fingerprinting is framed as a scientifically sound and incontrovertible tool to right precisely the inequities of the US justice system that underscore the conviction and incarceration of innocent people. For *The Innocence Project*, genetics both evidences and works against the confirmation biases of justice.

At the same time, the *Project* relies on assumptions of the evidentiary certainty of DNA testing. DNA fingerprinting is, of course, a distinct scientific arena from behavioural genetics and arguably far less subject to the indexical vagaries and complexities that converge in the latter. Yet it is a field that is also constituted of uncertainties, not only from a purely biological standpoint, but also as it articulates with other fields, including justice, anthropology or evolution, which in themselves are complex. In his 2013 *Slate* article 'Doubt and the Double Helix',[13] Jordan Ellenberg,[14] distils three points concerning the uncertainties of forensic DNA testing. First is that DNA tests involve comparison of markers between samples (which can be a singular comparitor or an assessment against a DNA database). The number of markers compared[15] will change the meanings that can be inferred from a 'match'. 'Matching' refers to statistical probabilities that the corresponding markers (Ellenberg cites the use of six)[16] from one sample to another describe the same person. Thus while popular wisdom is

---

– Not include a "sunset provision" or expiration date for post-conviction DNA access.
– Require states to preserve and account for biological evidence.
– Eliminate procedural bars to DNA testing (allow people to appeal orders denying DNA testing; explicitly exempt DNA-related motions from the restrictions that govern other post-conviction cases; mandate full, fair and prompt proceedings once a motion seeking testing is filed).
– Avoid creating an unfunded mandate, and instead provide the money to back up the new statute.
– Provide flexibility in where and how DNA testing is conducted.
(<http://www.innocenceproject.org/fix/DNA-Testing-Access.php>).

13 Ellenberg (2013) article is based on (and a review of) Schneps and Comez (2013) book: *Math on Trial: How Numbers Get Used and Abused in the Courtroom.*

14 Jordan Ellenberg is a Professor of Mathematics at the University of Wisconsin.

15 Likewise, the size of a comparitor data base will affect odds.

16 As Jones (10 April 2010) explains, the entire genome is not sequenced, but only a very small portion of it, in the UK, the norm of 11 markers used in the UK constitutes 'about one millionth of the total'.

that individuals have unique DNA profiles (and this may well be the case), the notion that DNA testing relies on or reads that uniqueness is a conceit.[17] The odds that more than one person might share a limited array of DNA markers may be very small, even infinitesimal, but they are not zero. Ellenberg's second point concerns the problem of *confirmation* bias (also termed the *prosecutor's fallacy*), that is, the assumption that the '1 in a billion' odds of matching another person's profile is an indicator of guilt.[18] As Ellenberg explains, the question of the odds of guilt is a different question than odds of innocence; and these different questions produce very different statistical ratios. Thus there may be a 1 in 5 million chance of a person who has nothing to do with a crime being found to be a 'DNA match' to a tested sample, but at the same time a 1 in 2 chance that a defendant (the 'guy who matched the sample') is innocent. In the typical courtroom, however, the DNA analysis presented is only the first of these two critical ratios. Even the inference that a DNA test may provide more plausible evidence that someone was *not* present at a crime (the key exculpatory claim used by *The Innocence Project*), than that they *were* present is problematised by this consideration. Third is the problem of cross-contamination; that is, as Ellenberg states, while a DNA test may strongly suggest that one is the owner of the tested sample, it does not provide conclusive proof of presence at a crime.[19] Finally, and perhaps most importantly, the foundational notion underpinning the perceived reliability of forensic DNA testing is that genetic/ genomic profiles of individuals are singular and unique (and, as a corollary that every cell of a person's body has but one unique genomic iteration). Yet, this understanding has been significantly ruptured; it is not only possible, but increasingly recognised as common, that individuals are composed of multiple genomes.[20] In this respect, the demands of forensic analysis would seem to be directly at odds with, and undermined by, knowledge of the complexities of biology.

As Ellenberg points out, DNA testing on its own does not deal in or produce certainties. And as detailed in its case files, *The Innocence Project* triangulates DNA

---

17 Having unique biological profile is distinct from being able to biologically measure or account for that uniqueness.

18 Jones (10 April 2010) cites the 2007 Nuffield Council on Bioethics report, *The Forensic Use of Bioinformation: Ethical Issues*, which states 'the *prosecutor's fallacy* has compromised the use of DNA evidence for a fair trial'. This fallacy is based on a misunderstanding of statistics.

19 See also, for example, Ogasogie (25 July 2013).

20 Thus, for example, in 2012, 'forensic scientists at the Washington State Patrol Crime Laboratory Division described how a saliva sample and a sperm sample from the same suspect in a sexual assault case didn't match' (Zimmer, 16 September 2013). Zimmer notes that multiple genomic profiles were found in autopsies of the brains of women who had been pregnant, in twins, and in recipients of bone marrow transplants.

testing where possible with other modes of counter-evidence. At the same time, the *Project* narrative powerfully reinforces DNA testing as a cornerstone in the pursuit, not so much of guilt (the bias of the justice system), but of reasonable doubt.[21] The danger, Ellenberg suggests, is that the cause of righting juridical malpractice can persuade against reasonable doubt in the science. Also eclipsed in this association are the social justice issues that arise with the aggregation of genetic data into a forensic database.[22] Whose DNA is going to be held, under what circumstances and to what consequence for those individuals – all of these questions accrue as much to bias as the verdicts overturned by the *Project* since they arise from the same context. The mission of *The Innocence Project* is not simply about misidentifications, but about the systemic flaws and injustices that underscore them. Herein is a dilemma: can genetic evidence arising from the myriad biases of the justice system, which is itself demonstrably compromised, provide an (uncompromised) corrective to that system's (unjust) consequences?

Crime and Genes as Meta-Narrative

[E]very contact you make, no matter how small, will leave a trace. (Calleigh, *CSI: Miami*, 'Out of Time')

The CIBA symposium and *The Innocence Project* represent different arenas of the cultural field. If CIBA represents a local imaginary constituted of experts in a more or less closed academic-scientific speech community, *The Innocence Project* constitutes the mid-level – intermediating localised frames of specialised expertise (legal-juridical, scientific) with political-economic institutions that constitute the justice system and public policy, and, in a limited way, by taking

---

21   The presumption of certainty concerning DNA testing frames the 2012 PBS *Frontline* documentary 'The Real CSI' (which happened to be screened in the UK as I was completing this book). The documentary critiques the ways in which the disciplines of forensics (blood spatter analysis, fingerprinting and so on) have developed without any system of scientific validation or accreditation. Thus, for example, the axiomatic presumption that fingerprints are unique has no scientific evidence to support it. DNA testing is cited as an antidote to the lack of scientificity in all other areas of forensics because it arose from the context of scientific research. There is no suggestion that science, itself, might also demonstrate problematic presumptions of cognitive bias. In making this argument, the documentary draws on the work of *The Innocence Project*.

22   McHale (2004) makes a comparable point about genetic data bases in the clinical context, with arising issues including consent, privacy and rights to bodily autonomy as well as potential discriminatory consequences for medical care, insurance and employment. See also, for example, Vikram (20 May 2013) on the retention of DNA from children in the police database the UK.

up a wider public presence through their website and citations in popular media fora. *CSI*, by contrast, represents the meta-level[23] – the arena of cultural mythology, trope and commonsense – the *popular* episteme.

Watching People Think

> CSI is a show about thinking and watching people think. (Anthony Zucker (*CSI* creator), cited in Cohan, p. 8)

The *CSI* franchise – internationally syndicated and including three separate shows: the original set in Las Vegas (2000–present); *CSI: Miami* (2002–12); and *CSI: NY* (2004–13) – is distinct for its dramatic setting at the scientific nexus of politics, policing and criminal jurisprudence. *CSI* is quintessentially an *epistemic* genre form; it is preoccupied with the nature of knowledge – its standpoint, its underpinning logics and its empirical base. At its narrative centre is theory-building. Dialogic exchange is constituted through interplay of evidentially organised action, interpretive narrativisation (also deployed visually through interpretive graphics) and analytic debate. *CSI* is also organised around a number of foundationalist premises (arguably most realised in the original) which are consistently reiterated in dialogue. First is that *evidence doesn't lie* – an axiom particularly associated with and asserted by the character of Gil Grissom. Second is that knowledge is not secure – the job of the *CSI* is, by corollary, to question what is known, including their own assumptions of knowledge. Yet the same time, knowledge is also understood as secur*able* – as evidence, of itself, and it will produce an authentic account. The central epistemic tension across shows (and its currency as a realist allegory) thus lies in the closing of the distance between evidence and narrative. 'Truth' is secured through a reciprocal reconciliation of evidentiary process (long visual sequences of painstaking methodology both at crime scenes and in the lab constitute a key trope of *CSI*) and narrational plausibility. Thus, at one and the same time, *CSI* demonstrates the contingency of evidence (each new piece demands a change of account), while maintaining the position that evidence is absolute, authoritative and objective. The arbitrariness of when and why the evidence gathering should stop (*why* should it stop?) takes on the currency of the latter proposition, while never resolving the former. In this context, it is *plausibility* that functions as truth, even as truth becomes an inference of persuasion. Passions, moreover, are understood to get in the way

---

23 I am drawing broadly here on distinctions between different levels of narrative set out in Epstein, Johnson and Steinberg (2000) in our analysis of parliamentary process and the transforming legal–cultural repertoires surrounding sexuality, equality politics and the age of consent in the UK.

of knowledge building (only the evidence matters), but knowers are portrayed always as passionate. These contrary tensions are distinctively personified in the character of Gil Grissom, who is at once profoundly engaged and autistically literal. Thus epistemic privilege (as *the* modality of public interest) is visually–narratively equated with the orientations of detail, of crime's microscopy, of trace. Both are constituted as the calculi of justice.

A third premise (as evoked in the epigrammatic quote above) is that everyone leaves something behind – the presumption of *trace*.[24] This is not simply that our bodies map the terrain of our presence in the world, but the converse trajectory – that these microfilaments of *someone's* presence can be traced back to their origins as/in specific persons. Thus *trace* evokes both a categorical (singular, unique) *identity* and a methodology of *identification*.[25]

Slam Dunk

Prosecutor: So you're bringing me junk science, is that it? You trying to put a loss in my column? I don't think so … Besides, I agree with your partner, Sullivan, on this one.

Horatio: He was here?

Prosecutor: Yeah. Helping me to put the husband away. Slam dunk. ('Out of Time', *CSI: Miami*, 2010)

As with *The Innocence Project*, the question of confirmation bias is generically central to the *CSI* imaginary and to its meta-level intermediation of genes, crime and justice. This question is played out explicitly as a plot device in the Season 8 episode of *CSI: Miami*, 'Out of Time'. This episode is interesting for the purposes of discussion here, in three particular respects. First is its portrayal – as a feature of its generic *raison d'être* – of the interface of forensics and law enforcement. In this respect, it is *CSI: Las Vegas*, that is emblematic of the show's critical epistemic centre, presenting a meditation on that interface as much as it does particular proceduro-moral puzzles within that frame.

---

24 The absence of trace at a crime scene therefore produces edifying dramatic tension for particular episodes, as it can be taken to suggest failures of method on the part of crime scene analysts (they missed something), and/or a sociopathically 'intelligent' (or media savvy) perpetrator – the 'accomplished' serial killer, or perpetrator who is well versed in crime procedurals.

25 In the series, 'Trace' as a department of the crime lab is distinguished from the DNA lab. However in the larger philosophical sense, both departments deal in trace evidence.

CSI: Miami by contrast, makes only the barest gestures toward forensic process, law and ethics; in its overarching frame of violent voyeurism directed both at female bodies (sunning, stripping, dancing – in a state of semi-permanent debauchery) and the quasi-militarised paraphernalia (Hummer vehicles, guns, drugs) of confrontation. 'Out of Time' thus stands out for the absence of this repertoire, favouring instead the tropes of *CSI: Las Vegas*.

A second point of distinction in this episode is its treatment of criminalistics through an *innocence project* narrative. The episode juxtaposes two parallel story lines surrounding three lead characters, Eric, Calleigh and Horatio. The first, taking place in the present day, focuses on Eric, one of the CSI team, who is grievously injured and found unconscious and near death. The second, told in flashback, is the murder investigation nine years earlier that brought these characters together and would lead to the formation of *CSI* as a new team of professionals and as a specialised new division of Miami police work.

It is important to note that *innocence project* narratives can be found frequently across all three *CSI* series (and indeed, across the genre). In 'The Execution of Catherine Willows' (*CSI* 2002) for example, a closed case – with a convict on death row – must be re-opened because of DNA evidence (which had been unavailable at the time of his conviction). In this instance, Willows must grapple not only with her own complicity in this injustice (she was the CSI on the original case), but must set aside the assumptions of guilt which she herself had validated.[26] What distinguishes the 'Out of Time' episode is its focus, as a point of narrative in and of itself, on the transformation of the forensic field, with the take-up of DNA testing.

The flashback plot centres on a miscarriage of justice – the wrong man has been pursued/arrested for murder. This is presented as not simply a case of mistaken identity, but a consequence of actively biased policing. This wrong man is pursued because the police are invested in, as the prosecutor states, 'put[ting him] away' – because he 'fits' the frame of plausible perpetrator. The underlying bias of this pursuit is established in two ways. First is the presentation of the police as ham-fisted buffoons who treat women badly – they engage in ugly and pointedly 'old fashioned' sexist banter as Calleigh is transferred into their team. Second, is their invested, impatient and incompetent rush to judgement. This is emblematised in the character of Detective Sullivan, Horatio's partner, who when faced with the body of a murdered woman (Amy Bowers) in her own home, immediately 'likes' the husband (Steve Bowers), an investment he and 'old school' cronies maintain despite explicit and, as the episode progresses,

---

26 As noted in an advance review in *Salon*, the Season 4 premiere of *The Good Wife* has an innocence project central narrative, and, going further, will feature *The Innocence Project* founder and defense attorney, Barry Scheck in a cameo role (Dunning, 26 September 2013).

mounting evidence to the contrary. That Amy Bowers is white and her husband is black ramps up the moral stakes, setting up a powerfully evocative meta-narrational 'reading' of the main plot. Much as the taunting of Calleigh stands in for a larger injustice, Sullivan's active pursuit of this specific wrong man, cannot but evoke the systemic pursuits – the 'guilt project' as it were – of racist policing and penal systems. The presumption of his guilt is thus framed as not incidental; not as a (statistically supported) inference that where a wife has died, there is a likelihood that the perpetrator is her husband; and not as simply the bias of one rogue police officer. Rather, the railroading of Bowers is played out on and as an explicitly racialised interpretive field.[27] Significantly, the injustices meted out to Calleigh and to Steve Bowers are consistently noticed (in the case of Bowers, this is entirely tacit) and quietly resisted/refused by Horatio.[28] The narrative dénouement is that both wrongs (the arrest of an innocent man and the racist–sexist biases of old fashioned policing) are remediated by the new forensic culture heralded by genetic testing.

What the Body Has to Say[29]

As Calleigh, Horatio and Eric[30] triangulate forensic evidence (blood spatter, soil), an alternative suspect, the landscaper Arnold Hollings (who is white) emerges. The blood evidence of the victim that they find on his body (a single drop in

---

27 For extended discussion of the racialised visual-interpretive field of criminal justice, and the ways in which the controlling mythologies of black masculinity at its centre played out in the case of Rodney King, see Butler (1993).

28 The episode is punctuated by moments of Horatio's active recognition – at times portrayed through his expression, gaze and minute gestures (a small nod, a glint) and at others through dialogue – of Calleigh's expertise and of the normative gendered climate disrespect that has framed her career thus far. At one point, amidst a discussion of evidence, she is suddenly smiling. Horatio asks her why. She tells him 'Because this is what I was hoping it would be like when I got here'. 'Me too', he replies. This recognition is meted out to Eric (Horatio's recognition of his potential), as well as to the wrongfully suspected Steve Bowers (almost entirely tacit) and of the sloppy incompetence and racial bias of his colleague (through 'tactful' withholding of comment).

29 The sequence:

Alex (the pathologist): Cop is convinced it's a domestic.

Horatio: I'd like to hear *what the body has to say*.

…

Alex: Where are you going with this Horatio?

Horatio: I'm just following the evidence.

(my emphasis)

30 Along with a fourth character, a disenchanted police officer, also scientifically inclined, who is on his last day with the force.

his nostril) is too small for traditional DNA analysis.[31] Three ensuing exchanges establish the moral as well as critical scientific capital of DNA analysis. In the first – an exchange between Horatio and the DA – Horatio frames the scientific and political stakes of their dilemma as mutually contingent and ethically zero-sum:

> Horatio: I have a friend in the Miami field office, she owes me a favor. If we don't find a way to stretch this DNA, an innocent man is gonna go to jail.

The subsequent exchange takes place in the scientifically advanced (and socially enlightened) FBI lab, staffed/led – in pointed contrast to the men's-room culture of the Miami police – with two female agents:

> Horatio: If we don't find a way to stretch this DNA, an innocent man is gonna go to jail.
>
> Agent: Well, couldn't have caught that serial bomber without your expertise, Detective.
>
> Horatio: It's the least I could do. I put a reference sample from our victim in here, too. But first step is to multiply the DNA. Agent Boa Vista, would you run a PCR on these samples? Thank you.
>
> Agent: Used to be we needed a blood sample the size of a quarter to run DNA. Now it's the equivalent of a few skin cells.
>
> Horatio: Progress marches on, huh?
>
> Agent: Yeah. Magic. DNA for you and DNA for the defense.
>
> Horatio: If it matches our victim.
>
> Agent: All right. This is the DNA profile for your victim. This is the unknown sample. *Odds that two people have the same profile are about one in a billion.*
>
> *(subtle crescendo of the musical soundtrack climaxing on the statement)*
>
> Agent: The blood is a match.

---

31 The suspect has inhaled the blood spatter of his victim. That Horatio notices this rather unlikely *deus ex machina* plot development reinforces the distinction of ethical policing (derived from meticulous orientation to detail and science).

Horatio: Arnold Hollings murdered Amy Bowers. Open-and-shut.

The citation in this exchange, of '1 in a billion odds' evidences powerfully the intermediation of criminalistic science, juridical process and cultural mythology. Here the *prosecutor's fallacy* is at once amplified into a metanarrational truism, even as it lends a generic *realism* to the phantasmic imaginary of television. Furthermore, the contrast of settings, of personnel, of values, are signalled by explicit moral freighting of DNA testing: it is 'magic'; it is 'progress'; it is an unequivocal 'match'; it is 'open and shut'. These associations mean that Horatio can invoke a virtually identical declarative – his 'Open and shut' to the prosecutor's earlier 'Slam dunk' – but only the latter is figured as brutal and wrong. DNA testing thus elides into justice, not simply in the forensic–juridical sense, but as *social* justice.

In the logic of this episode, Horatio and Calleigh (and later Eric) represent a new culture of policing. One oriented to detail, to observation, to science. All break out of a moribund and corrupt system. This is signalled by Calleigh's technical expertise and articulacy in ballistics (a hypermasculine terrain), Eric's heretofore unrecognised potential (it is Horatio who recognises that he is under-utilised as a salvage worker and begins to train him informally) and, most importantly, by Horatio's interest in, and privileged access to, state of the art DNA analysis. It is also signalled by the episode's meta-level allegory of racial (in)justice.

The final sequence sees a montage of scenes (between the DA office and elements of the reconstructed crime), backed by a voice over from the court proceedings where Hollings is charged with murder. As part of this montage, and against a soundtrack that is both melancholy and thoughtful, we see the exonerated Bowers released from his cell. As he is set free, his eyes meet Horatio's. Each gives a barely perceptible nod. What they each acutely recognise – what has been remedied here – is not simply the immediate wrong to an individual; it is also the larger injustice. This is indeed, the innocence project.

Open and Shut

There's only so much reasonable doubt we can genetically engineer away. (Ellenberg 2013)

I would like to conclude with two points of observation. First, concerns the question of *episteme* – that is, the conditions of possibility of knowledge. What is demonstrated by these three case studies is their participation in a circulating ecology of meaning in which scientific, socio-political and cultural discourses of crime, criminalistics and justice are inter-referential, even as they occupy distinctive knowledge niches, address different constituencies of knowers and constitute distinct modes of public capital. All three articulate normotic[32] assumptions concerning genetics and crime (assumptions of genomic uniqueness, of trace, of definitive genetic methodology, even of predictive potentiality) even as these investments differ in standpoint, forum or focus. At the same time, in this triangulated context, the gene also emerges as a locus of contestation in two respects. As a matter of science, there are competing paradigms concerning the suture of genes to crime and to identity, and these are emergent in both the diagnostic as well as jurisprudential frames – to the extent that recent developments in genomics (i.e. findings that bodies are normatively constituted of multiple genomic profiles) suggest the wholesale unravelling of the place of genetics in the justice system overall. Yet on the terrain of justice, the gene as both central referent and framing discourse also seems at its most secure. This is in part, I would suggest, a consequence of the emotional freighting of justice as a cause – and more particularly, as a terrain of *ethical* facts. To put it another way, justice in all three of these case studies constitutes not only a *way* of knowledge – but an effective, justifying *persuasion*. The imperatives of justice as a philosophical-ethical standpoint can powerfully sediment an investment in certainty. This is perhaps nowhere more evident than in the context of a justice *system* given the potentially grievous consequences for human lives and for the rule of law in and of itself, if justice is arbitrary. The progress of science can as much rupture as it can solidify the standards of criminal investigation and reliability of evidence. Interestingly (and perhaps unsurprisingly) it is only in the phantasmic space of *CSI* that the counter-possibility – *uncertainty* as justice, indeed uncertainty as *good science* – is entertained. And even there, that possibility is consistently displaced, transmuted to a residuum of unease that attaches to (but does not override) its pre-eminent trope of gene(tics) – the 'positive match'. Thus, while it may be the case, as Ellenberg suggests in the quote, cited earlier in this chapter, that 'there is only so much doubt that we can genetically engineer away', it is also the case that stakes of certainty (as distinct from the facts) constitute a powerful current of persuasion.

---

32 As noted earlier in this book, I am drawing on (and also departing from) Bollas' (1987) understanding of 'normotic' in the psychoanalytic context. Normotic assumptions or values, as a cultural description, refer to the invested repudiation of what is subjective (feeling, the body) in favour of an idealised ultimate rationality and a concomitant investment in (and assumption of) control.

Chapter 6
Beggars and Choosers:
Genes and the Neoliberal Subject

Behold I tell you a mystery. We shall not all sleep but we shall all be changed, in a moment, in the blinking of an eye, at the last trumpet. The trumpet shall sound. And the dead shall be raised, incorruptible. And we shall be changed.

(I Corinthians 15: 51–2)

Introduction

I am not the first person to suggest the interest of science fiction as a source (as well as interesting object) of social theory. It is a genre whose *raison d'être* is critical engagement not only with social issues, but with the detailed and complex everyday ways in which a social order may be lived out, the feeling structures that support and are produced by divergent political orders, and the possibilities of (and foreclosures on) individual and collective agency and social change. Science fiction has also notably engaged in speculative meditations on biopower and bioethics and on the reciprocal constitution of knowledge and power. In this chapter, I would like to consider what science fiction might contribute not only to a critical understanding of particular bioethical issues – in this instance, the convergence of genetics and neoliberal politics – but to a re-imagining of the bioethics field itself.

As Stacey (2010) has noted, science fictions of the gene abound across the cinematic landscape, deployed in both utopian and dystopian interpretive renderings: as allegory, trope, metaphor, and referent; as devices (or objects) of social commentary; in realist verisimilitude; and in magical fabulation. As Stacy argues, this cinematic landscape provides a potent signpost not only of the 'genetic imaginary', but of the wider cultural zeitgeist. In this context, genes and cinematic convention constitute inter-referential frames of intelligibility for bodies as they articulate with politics, technology, identity and intersubjective relations. The treatment of relations of 'difference' – of gender, race and class; of what is 'alien' and what is 'familiar' – have constituted particular preoccupations. These considerations also, of course, extend to the literary context where the gene has long constituted a dominant motif.[1] Genetic engineering (and associated reproductive technologies) provided a significant

---

1 This spans a science fiction tradition from Huxley to Herbert.

point of projective social analysis during the 1980–90s renaissance of feminist publishing. Feminist science fiction flourished in this period, with feminist publishers like The Women's Press in the UK having dedicated genre-defined lists (e.g. feminist science fiction, feminist detective fiction), and mainstream publishers engaging in high profile distribution and promotion of feminist science fiction authors from Marion Zimmer Bradley (Tor), to Sherri Tepper (Bantam), to Marge Piercy (Knopf; Penguin), and Margaret Atwood (Anchor Books). Distinct from the wider genre (which have tended to frame genetics either itself as dystopian[2] or as co-opted into the service of a dystopian political-economic governmentality),[3] some feminist science fiction novels took up a fully realised utopian 'read' of recombinant genetics. Joan Slonczewski's (1986) *A Door Into Ocean* is perhaps most notable for its counterposing of a feminist bioethic of 'life shaping' (the mode of using 'life-stuff' to manipulate life and parthenogenetic reproduction) against the death-seeking (genocidal) militarised science of the patriarchy.[4] The majority offer ambivalent reads on genetification as a social condition and as an organising logic of (and at times against) regimes of embodied difference into power.

This chapter will examine the *Beggars* trilogy by feminist science fiction writer Nancy Kress, concentrating particularly on the first novel, *Beggars in Spain*. The *Beggars* novels offer a potent meditation on two particularly salient points of contemporary rupture and cultural contestation: the advent of human genetic engineering, and sleep – or, more specifically, the prospect of a sleepless society. Kress' subject matter, set in a not so distant future USA, is the genetic engineering of a new class of 'sleepless' – the ideal worker-citizens of a new cosmopolitan (and post-fossil fuel) world order. My interest here is, in part, the place of science fiction as both a reflection of and a commentary on the feeling–knowledge regimes of a particular cultural moment, what might be termed the *cultural episteme* or, following Foucault's (1966) early theorisation, the conditions of possibility of knowledge. Kress' works, characteristic of the genre, both anticipate and theorise the tendencies of the social order that produced them. They articulate realism with excess and the absurd, insight with counterintuitive devices of plot, asking the reader not only to consider, but in a sense to 'live', at times *reductio ad absurdum*, the underpinning logics and futural tendencies embedded in current scientific and political practices. Sleeplessness, in the Kress works, offers a symbolic apotheosis – an ultimate convergence – of real

---

2 For example, Herbert (1976).

3 For example, as a corrupt medium of totalitarian governmentality or corporate greed (Huxley 1932; Crichton 1993; 1997; Piercy 1976; 1993).

4 Sherri Tepper's earlier (1980) *The Gate to Women's Country* provided a comparable ethical parable as (implicitly genetified) patriarchal traits are systematically bred out in the pursuit of a life-affirming matrifocal society.

contemporary trends toward a 24/7 society and of real social transformations generated not only by new techniques of genetic manipulation, but the genetification of the social field itself. The *Beggars* world offers, as we shall see, an allegory of class transformation and embodied distinction in the wake of technological and biopolitical revolution. The novels foreground cultural ambivalences that are distinctly bioethical, elaborating both dystopic and utopic projective fantasies about bodies at the interstices of discourse, technology, and social order. In so doing, the *Beggars* fantasy – like that of Stephanie Yanchinski in *Setting Genes to Work* – leads us to the centre of the 'crisis of bioethics' itself.

This chapter begins with the proposition that bioethics is at an impasse.[5] While there is certainly no singular approach to bioethical thought, as the contributors to Murray and Holmes (2009) suggest, the field has nonetheless come to be dominated by a legalistic preoccupation with regulatory structures, an affective investment in consensus, and an affirmative orientation to scientific innovation. Indeed, at times there seems to be a veritable bioethics mill, directed at 'controversy' and perceived 'extremity', and aimed less to critically interrogate than to slough off the contestation and uncertainty that can (and, I would suggest, *should*) attend biotechnological transformations of the social world.

In forging this analysis, I aim to move my consideration of the gene onto the bioethics terrain and, in so doing, to contribute to the bioethics field in two ways. First, I want to engage with bioethics as a genre of representation and a framework of engagement, as well as a field of knowledge. In this context, I aim to incorporate an element into a bioethical critical standpoint that is not usually part of the discourse – that is, the question of affect. I want to ask what happens to bioethical discourse when it is concerned with what might be termed the 'feeling regimes' or 'feeling structures' of knowledge, body practices, modes of governance, and science. An analysis of science fiction offers a useful (though not the only) way into such a consideration. This is in part because of the necessary and explicit articulation of feeling and knowledge that characterises narrative as opposed to traditional academic (or even of popular scientific) genres of writing or regulatory modes of discourse.

Second, I wish to look at a particular example of science fiction fantasy to consider the ways in which it both re-presents and critically assesses contemporary body-political anxieties and scientific trends. Here I treat the Kress works as an affectively infused meditation on the confluence of neoliberal body politics and genetic engineering. The novels, I shall argue, by their strategic counterintuitive understandings of the body, suggest the intrinsically destructive dimensions of both neoliberalism and genetics – indeed, showing that each is an artefact of the other. At the same time, the Kress works also elaborate the phantasmatic,

---

5 For further discussion of the 'crisis in bioethics', see Shildrick (1997); Ettorre (2006); Murray and Holmes (eds) (2009).

seductive underpinnings of this confluence: the desires (and the denials) in play, desires anchored in both *commonsense* and *common-emotion* – those unremarked upon 'already knowns' and preferred affective orientations that constitute everyday life. It is this insight that allows us to leap from the fictional to the real, to consider the world we live in from an affectively inflected bioethical standpoint and thereby to problematise not only what disturbs, but what is comfortable and consoling. This is a bioethics that is not primarily oriented to consensus-building or to a simply affirmative, even if regulatory, approach to scientific innovation. Rather, it suggests a bioethics that has something of the character of McClintock's idea of a science predicated on a 'feeling for the organism'[6] – and by corollary, as a feeling enterprise.

Behold I Tell You a Mystery

Over the course of the early- to mid-1990s, Nancy Kress published the *Beggars* trilogy, beginning with *Beggars in Spain* in 1993, *Beggars and Choosers* in 1994, and *Beggars Ride* in 1996. This trilogy, as I have noted above, emerged in a context of a burgeoning of dedicated publication of feminist science fiction. It took its place alongside a plethora of fictionalised and scholarly assessments of genetics, of competing bioethics, of competing speculative social orders, and indeed of competing feminist paradigms. In both contexts, the relationship of the political to the embodied, and of feeling to knowledge, provided the foundations of what Haran (2003) has termed the 'utopian impulse' of feminist theory itself. As we shall see, the core themes of the Kress trilogy anticipated and continue to have contemporaneous bioethical resonance at a number of levels. The focus on sleep and sleeplessness provides a pointed metaphor for a society in which 24/7 awakeness is technologically possible. This literary device refers implicitly to a contemporary social condition in which we increasingly see 24/7 expectations placed on the functioning of social institutions and human bodies. The ethical values embedded in 'sleeplessness' and the instrumental manipulation of human beings through genetics play into two linked popular and bioethical anxieties: on the one hand, the boundaries of *humanness* and on the other, the obligations of *humaneness*. Both are troubled themes that have stood at the heart of science fiction explorations, applied to the speculative consequences of scientific hubris, rampant capitalism, and authoritarian governmentality. They are points of ethical meditation, highlighting questions of social responsibility, political dissent, and the possibilities of justice and transformation.

---

6 See Fox Keller (1984).

Beggars in Spain

Beggars in Spain opens with a scene that takes place in an IVF-genetics clinic set in a not-so-distant future USA. Roger Camden, an extraordinarily successful business tycoon (described as a 'data-atoll investor') and, himself, a brutal embodiment of hyper-productivity, is negotiating to produce a daughter with the genetic enhancement of sleeplessness, a trait that is part of a secret experiment. His wife Elizabeth, who sits with him, is a cowed, sullen, and yet resistant figure who we later learn has a drinking problem directly as a response to Roger's ruthlessness and his inexhaustible drive. Indeed, Roger is, himself, virtually sleepless (and regards this as his highest virtue). By accident, Elizabeth conceives twins, one enhanced through the IVF-genetic process and one, naturally. The twins are Leisha and Alice – sleepless and sleeper, respectively. This sets up two counterpoints of intimate tension that are followed throughout the rest of the novel (and also in increasingly complex forms throughout the trilogy). These are, firstly, tensions between Roger and Elizabeth which emblematise a clash of gendered neoliberal values – the imperialist machismo of enterprise and acceleration (culture) versus the passive and entropic properties of domesticity (nature). An alternative tension is set out in the competing femininities of Leisha and Alice, embodying the contrasting values of *fast time* versus *slow time* and troubling the intersecting questions of what is human and what is humane.

We are further told that the social world in which this experiment to genetically engineer sleeplessness takes place is one already transformed by 'Y-energy', which has replaced fossil fuels. Y-energy is a form of cheap, ubiquitous, and limitless energy invented by Kenzo Yagai. While the novel never specifies the particulars of Y-energy production, we are given to understand that it is essentially the realised achievement of a counterintuitive proposition: the possibility of perpetual motion. This Y-energy driven society has also been transformed by 'Yagaism', a political philosophy of meritocratic social order where individual excellence and industriousness are the basis of the social contract – itself understood in marketised terms as 'trade'. Both Y-energy and Yagaism constitute a new world order in which the utilitarian and anti-entropic values of neoliberalism, articulated as a collectivist philosophy, reach their apogee (and perhaps their own undoing).

The birth of Leisha and the other sleepless children presages a radical rupture in a world in which genetic enhancement has already been assimilated as a normative practice in the service of social privilege (and 'good' social order). The appearance of the sleepless heralds what unfolds as a dramatic transformation of class hierarchy and a new form of racialised distinction. The sleepless themselves are posited as unusually beautiful (with few exceptions), possessing an intelligence beyond the range of 'natural' norms and as preternaturally industrious. As the plot unfolds, it further emerges that the sleepless are extraordinarily bodily efficient – sleeplessness means that they virtually do not age. Sleeplessness furthermore turns out to be a dominant genetic trait, thus creating a new 'natural order' as sleepless adults have sleepless children and constitute an increasingly separate people with a distinct culture accruing directly from the sleepless biological trait. The sleepless, moreover, exactly embody the ideals of Yagaism. They are in perpetual productive motion with virtually no material cost or embodied friction. They are embodied paradigms of individual excellence: innovative; supremely industrious; and innately imbued with ideal affect – being unflaggingly sociable, positive, and fulfilled in service.

As the novel proceeds, a new world order is produced with distinctions between the sleepless; the 'donkeys' – those who are genetically enhanced but not sleepless – who monopolise most of the labour (a professional order of labour as robotics take care of the drudgery); and the 'Beggars' (also referred to as 'Livers'), the genetically unenhanced majority who do no work at all, but live a life of 'low' hedonistic pursuits, with an overweening sense of entitlement and who, in Yagaist terms, offer nothing in the contractual trade of a meritocratic civil society. In amidst this new order, troubled (and troubling) figures appear, most notably Jennifer Sharifi, a 'flawed sleepless' who emblematises the irreconcilable tensions and hostilities brewing within the new order. As we shall see, Jennifer comes to stand as an ambivalent ethical commentary on the potentialities and foreclosures of neoliberal political economies. Her paranoia and fanaticism signal an underpinning disquietude, a fatal flaw both of desire and disaster.

We Shall Not All Sleep

> Ong Smiled. "Appearance factors are the easiest to achieve, as I'm sure you already know" …
>
> "Good enough," Camden said. "The full array of corrections for any potential gene-linked health problem, of course."
>
> "Of course," Dr Ong said …
>
> "And," Camden said, "no need to sleep."
>
> (*Beggars in Spain*, Kress 1993, 4)

In her 2002 study, Elizabeth Ettorre discussed the ways in which values of capital (human, social, economic) accrue to genetics and genetic modification. This represents a long-standing theme of feminist and other critical assessments of genetic science, including historical analyses of its antecedent developments within nineteenth- and twentieth-century eugenics.[7] The value of capital (positive or negative) attaches not only to attributed properties of bodies (genetic traits, lineages, relationships), but also to the scientific, economic, and political enterprises surrounding genetic practices, as well as to the growing hegemony of genetic ideas and ideologies. This theme is prominently articulated in the *Beggars* trilogy. As evidenced in the passage above, taken from the stage-setting scene at the start of *Beggars in Spain*, this is a future US in which genetic enhancements (and genetics *as* enhancement) are both achieved and normative. Ong's 'of course' evokes genetic modification not only as a taken-for-granted everyday underpinning of social life, but also as a compact of understanding between two men, united both by superordinate social position as well as knowledges (both know what is *not*, as well as what *is*, widely known). The exchange between Ong and Camden invokes seductive utopian tropes characteristic not only of genetic engineering-themed science fiction, but also of the public relations often attached to genetics itself.[8] These are the notions that genetic modification can beautify the body surface, even as it can improve the elegant efficiency of an organism. That such precision is possible – though perhaps not ethical – constitutes much of the hype that has surrounded genetic science. Such notions, one might argue, stand at the fulcrum between seductive, marketised fictions and projective fact.

Genetic capital in the *Beggars* world, furthermore, is part of a circular economy that is foundational to modern (including late-modern) capitalism: it is the purchase of the already well-off and privileged, and it produces new indices of privilege. Interestingly, the genetic modification enterprise in itself is posited as a comfortable extension of the political economic values of the contemporary capitalist world. It is a projective future in which the tensions and debates about genetics have been resolved and successfully assimilated: Ong's 'of course' signifying a commonsense understanding not only of the distance we have come from such times, but of the 'unjustified' paranoia that characterised the troubled early reception of genetic modification practices. In the *Beggars* world, genetic modification is status quo; introducing new embellishments, yes, but offering no breach within or of the social order. It is rather the engineering of the specific trait of sleeplessness that is offered up as the point not only of rupture, but of radical schism.

---

7 See Kirkup et al. (1999); Steinberg (1997) for summary discussion of key periods of feminist debate on reproductive and genetic engineering that form the immediate theoretical context of the Kress works.

8 See, for example, Haran et al. (2007).

Embodied Capital and Entropic Life

"Without that energy expenditure [REM sleep], nonsleep cerebrums save the wear-and-tear and do better at coordinating real-life input, thus greater intelligence and problem solving … suppress REM sleep and people don't get depressed. The nonsleep kids are cheerful, outgoing … joyous" …

"At what cost?" Mrs Camden said. She held her neck rigid, but the corners of her jaw worked.

"No cost. No negative side effects at all."
(Dr Susan Melling and Elizabeth Camden in *Beggars in Spain*, Kress 1993, 12–13)

The *Beggars* world is premised on a set of interestingly counter-intuitive reversals concerning the functioning of the body and the entropic effects of time and energy expenditure. Entropy describes the process whereby useful energy is transmuted into non-usable form. It is, in other words, the process of decay, decline, disarray, and disengagement. Ageing is entropy. Stress is entropic. Friction is the guarantor that there will be no possibility of perpetual motion. In the *Beggars* world, the functions of sleep and sleeplessness are both reversed and, significantly, articulated through a distinctly neoliberal moral discourse concerning the nexus of bodies, work, and value. From the first to the final book, the hegemonic view of sleep (which is both problematised and romanticised simultaneously), is constituted as a failing. It is posited in scientific terms as the literal empirical basis for, and direct cause of, mortality. It is understood pervasively as the wasting of energy – the using up of the body – both in terms of industry (the body unavailable for productive work) and in terms of the very viability of the organism. Sleep, in other words, is the ultimate entropic trait of the living world. Sleep is also understood as that which is inelegant and ugly about the body: the signifier and material harbinger of the body's propensity to decline; to social disengagement; to parasitism – sleep, as we are repeatedly told, is the *Beggars'* trait. As suggested in the above passage – a sentiment that is reiterated by various characters throughout the trilogy – the sleeper is, by reason of this trait, predisposed to inferior intellect, to entropic, destructive emotions, and to compromised social and economic utility.

The sleepless body, by contrast, is constituted as anti-entropic. Sleeplessness is presented as the literal – and importantly, a profoundly *desired* – embodiment of the values not only of industry, but of neoliberal governmentality, both in terms of physical efficiency and affective orientation. The sleepless, it is oft repeated, are by nature industrious, invested in service, and optimistic. They are naturally excellent, superior, and embody materialisations of a conjoined economic and moral utility that supersedes the merely enhanced. The sleepless are offered as exemplars both of a nostalgically individualistic Protestant middle-class work ethic and an imperialistic liberalism of contemporary forms of what may perhaps be usefully termed 'cosmopolitan globalisation' (Johnson and Steinberg 2004). The sleepless, moreover, are constructed as frictionless, both in terms of bodily function and affectivity. They, like Y-energy, emblematise one of the most potent fantasies of modern life: the possibility of inexhaustible self-renewal. Both are figurations that stand in as consoling phantasmatic projections – and as impossible symbolic resolutions – to the profound contemporaneous crises that have accrued to the inevitable drying up of a fossil fuel-driven political economy. As self-renewing resources, sleepless and Y-energy both embody the surface fantasy[9] of perpetual motion and the powerful normotic[10] phantasy, central to religious discourses not limited to Western forms, that life itself may be transubstantiated, transformed, absolutely known, and live on eternally.

And the Dead Shall Be Raised

Beggars in Spain can be read as a parable of class revolution, and one that is constituted on and through a racialised field. Marx (and Marxist thought) has theorised the ways in which industrial capitalism became the bedrock of a radical reconfiguration of the social order. Thompson (1991), for example, has argued that the industrialisation of labour and early capitalist forms of exchange saw both the invention of the working class as alienated labour and of a new middle class who owned the means of production and profited from the labour of others. The narrative structure of the *Beggars* trilogy follows the dominant narrative structure of such historical accounts of class formation and the rise of Victorian modernity and Victorian values. Indeed, the sleepless can be seen, in this light, as paradigmatic modern workers and citizens. At the same time, Kress' projective fantasy of a genetic modification for sleeplessness is also understood to be inassimilable into the modernist values and social order

---

9 'Fantasy' is used here in the denotative sense of (more or less conscious) imaginative projection. I use the psychoanalytic term 'phantasy' to refer to unconscious, inchoate desires and fears that can underpin such projections.

10 See Bollas (1987) for discussion of 'normotic' in the psychoanalytic context. As noted earlier in this book, 'Normotic illness' is, in essence, the transformation of the self into an ideal object, a virtual automaton of 'normality', totalisingly invested in facts, utility, knowability, and rational control. Taken as a cultural description, a 'normotic phantasy' might describe the invested repudiation of what is subjective (feeling, the body) in favour of an idealised ultimate rationality. As such, it can be suggested that normotic phantasy constitutes the cultural 'unconscious' of modernity and modern science, with their principled idealisation of rational utilitarianism, objectivity (the repudiation of feeling and body), and law.

that produced it. In the first part of *Beggars in Spain*, the rupture is ominous and apocalyptic:

> Leisha, you're a different kind of person entirely. More evolutionarily fit, not only to survive but to prevail. Those other objects of hatred you cite—they were all powerless in their societies. They occupied inferior positions ... Every Sleepless is making superb grades, none have psychological problems, all are healthy and most of you aren't even adults yet. How much hatred do you think you're going to encounter once you hit the high-stakes world of finance and business and scare endowed chairs and national politics?
> (Stuart Sutter [Leisha's 'sleeper' college boyfriend] in *Beggars in Spain*, Kress 1993, 51)

The first half of the novel witnesses the Sleepless subjected to Nuremberg styled laws, excluding them from all manner of sport, work and community.[11] They are persecuted, subject to physical attacks and treated as mutant, repudiated Others – their perceived monstrosity to the sleeper majority is based on and articulated through a distinctly racialised repertoire of envy. Indeed, sleeplessness as a biogenetic trait comes to stand in for and refer to the panoply of reviled Others that have accrued to modern racial taxonomisations. Interestingly, though only given a passing reference – the overwhelming majority of characters at the centre of the initial narrative are white, with the notable exception of Jennifer Sharifi – the *Beggars* world is posited as a post-racist world. This is as an extension of a more generalised egalitarianism that is attributed not only to international relations but, in many respects, gender relations. Indeed, the *Beggars* society would seem to embody a variant of what Lemke (2004) identified as genetics' *regime of equality*. Though the novel never explicitly engages with the forces (genetic or otherwise) that have caused the de-suture between gender and social rank, the markedly and conventionally unequal relationship between Roger and Elizabeth in this context signifies both the residual tendencies of a previous (pre-genetified) order and, at the same time, the latent (masculine–imperialist) tendencies of a post-genetics world. As Haran (2003) has noted, references to National Socialism and its holocaustic racial ideology are a common thread in the dystopian currents of post- (and in some cases contemporaneous with)[12] World War II science fiction. The hatred to which the Sleepless are subjected is

---

11 There is a resonance here to the total ban on drug (and other) induced advantage in sports competition, as in the Lance Armstrong case. Though in a projective world where privilege is universally secured through engineered enhancement, the distinction of 'induced' and therefore unfair advantage would seem to be inevitably problematised. And indeed, in the current, real world of sport science and performance, the difficulty of making this distinction continues to arise.

12 See, for example, Burdekin (1987 [1937]).

not only cast in this light, but the engineering of sleeplessness itself is construed as unleashing and revivifying tensions of the previous racial order. Interestingly, Kress explores the intersubjective relations of racial hatred though the metaphor of *genetic* difference. In both contexts, racial hatred is understood as an intrinsic tendency that directly accrues to and follows from racial distinction. There is an implication that genetic science has both undermined the scientific-biological validity of 'race' (in its modern taxonomic forms) – hence the post-racism of a genetified world – while at the same time having the power to (re)introduce it, this time in an empirically 'valid' form. That what is engineered (sleeplessness) is counterintuitive to all that is currently known about biology and entropy, sets up a compellingly ambivalent assessment of the underpinning politics and social implications of genetic science.[13] It is in this 'Nuremberg' period, as I shall discuss below, that a schism among the Sleepless is produced and the most profound dystopic elements of the narrative (and the theoretical analysis it forges) is set out.

Fast Time

The latter half of *Beggars in Spain* witnesses a radical shift in the social order and the emergence of a new status quo. The explicitly classed languages of the new structures emerging from the assimilation of sleeplessness are interesting to consider:

> The United States was a three tiered society now: the have-nots, who by the mysterious hedonistic opiate of the Philosophy of Genuine Living had become the recipients of the gift of leisure. Livers, eighty percent of the population, had shed the work ethic for a gaudy populous version of the older aristocratic ethic: the fortunate do not have to work. Above them—or below—were the donkeys, genetically enhanced sleepers who ran the economy and the political machinery, as dictated by, and in exchange for, the lordly votes of the new leisure class. Donkeys managed, their robots labored. Finally, the Sleepless, nearly all of

---

13 It is one thing to fantasise a perfection of the human body in an intuitively and essentially desirable way; the fact that the fantasy here is counter-intuitive and by no means completely desirable, focuses the argument on the social, rather than the personal implications. 'Do not', Kress is telling us, 'imagine some obviously beneficial genetically-modified disease-free body; consider instead this very equivocal product and its implications'. The rupture of multiple commonsenses concerning body–capital– governmentality, is metaphorically figured through a continuous interplay of 'essentially, intuitively desireable' (the normative enhanced beauty of most of the sleepless) and 'equivocal' (sleeplessness is not only counter-intuitive to 'real' bodies, the *Beggars* story arc follows the *sleepless* into its second-generation engineering of *super-sleepless* with its main side effect – an unbeautiful body).

whom were invisible in Sanctuary anyway, were disregarded by Livers, if not by donkeys. All of it, the entire trefoil organization—id, ego, and superego, some wit had labelled it sardonically—was underwritten by cheap ubiquitous Y-energy. (Kress 1993, 238)

The obviously Marxian references invoked here (populism, aristocratic ethic, leisure, opiate of the masses) places the new society both within and in contrast to the class categories of (post-)industrial capitalism. If genetic modification is a marker of middle-class distinction, sleeplessness is the trait that embodies the new middle-class ego-ideal as well as its new aristocracy – its superego. While a schismatic hierarchical ordering of social life remains intact (and familiarly so), the character of classed existence has radically shifted. Leisure linked with plenty, once a marker of privilege, becomes a marker of dispossession; hence, a 'new poverty', but this time of utility – the 'Livers' are a 'useless', infantilised, and parasitic majority. They embody the trajectories of life defined by consumption, in a radical breach from the means and mode of production. No longer alienated bodies of toil for the profit of others, they are the excessive bodies of gross consumption and infantile regression at the expense of others. Powered by ubiquitous self-renewing energy (both Y- and sleepless), the situation of the 'Liver'/Beggar is a destitution defined by plenty. The remade working class do not work but anodise themselves with bread and circuses and an opiate mythology that they are better off than donkeys, who do or manage nearly all the work.

The resonances of this portrait with many aspects of contemporary life, or at least with pre-austerity Western Democracies, are unmistakable. Late neoliberalism witnessed just such a transformation of labouring life, with the monopolisation of work, rather than leisure, increasingly serving as a marker of middle-class status and a day-to-day reality.[14] Even if bodies do still inconveniently sleep, the advent of 24/7 life has a clear technological reality, based materially in digital communications, but also conscripted as a cultural ideal increasingly ascribed to all aspects of civil society; blurring boundaries between public and private, between night and day, between action and its contemplation. The neoliberal affect, invested in perpetually climbing standards of merit and excellence, in the superordinate value competition and global markets, and in inexhaustibility as both a premise and an ideal that can be attributed to our resources, is increasingly a normative mode of institutional rationalisation, applied as much to public services and educational reform as to

---

14 The situation post-2008, has seen a shrinkage and bottoming out of significant tranches of middle-class life (in the USA, Britain and Europe) in terms of pay, conditions, educational advantage and opportunity.

business, as much to modes of institutional governance as to personal habits in the privacy of the home.[15]

In the *Beggars* narrative, the beautiful body of Leisha, her triumph of brilliance and empathy, is the emotive linchpin of this bioethical projection, and of its utopian purchase. She is at once a familiar fetish and a paradigmatically compelling ego-ideal. Yet there is also a darker undercurrent here, subtly elaborated in the counterintuitive propositions of Y- and sleepless energy. This is an undercurrent that bespeaks the realities of bodies as, in actual fact, *exhaustible* resources, of the exhaustibility of all resources that support life and of 'exhaustibility' as a characteristic of life itself. These are realities that, by their very nature, can have no purchase, no place, no means to realise a neoliberal utopia. An economy of finite and declining energy inevitably attaches both to political ecology and to the entropic properties of living bodies. The neoliberal ethic is, thus, one that the body will always fail. At the same time, the *Beggars* trilogy's emphasis on the superlative elegance of genetically-enhanced and sleepless bodies highlights and, indeed, invests in the profound seductions of what is intrinsically an impossible ethic. Recombinant genetics emerges, at one and the same time, as a realist referent to ground a transcendent fantasy and as a magical phantasm, by which means a materially impossible social and bodily renaissance will be accomplished. In the *Beggars* world, both sleeplessness and genetic enhancement stand in as reciprocal metaphors and as intertwined phantasmatic projections, narcissistic and normotic – or what Freud described as the 'phantasies of action'.[16]

Incorruptible

> Jennifer disturbed her. Not for the obvious reasons she disturbed Tony and Richard and Jack: the long dark hair, the tall, slim body in shorts and halter. Jennifer didn't laugh. Leisha had never met a sleepless who didn't laugh, nor one who said so little, with such deliberate casualness. (Kress 1993, 43)

As I have noted above, the trait of sleeplessness connotes a classed ideal of embodied utility, but at the same time, particularly as set out in the first half of *Beggars in Spain*, it emblematises a rearticulated racial order in which genetics emerges as a 'new-racism'.[17] In the *Beggars* world, the superhuman can stand

---

15 For extended discussion of these transformations as they have played out in the British pre-austerity context, see Johnson and Steinberg (2004).

16 The psychoanalytic concept, 'phantasy of action' refers to the projective recovery of the phallus (power in the social world).

17 Barker (1983).

in, simultaneously, for both ideal and repudiated. The dramatic tension arising from the ambivalent effects of genetification, generally, and the modification of sleeplessness, specifically, emerges as a rupture, not only between sleepless and sleeper, but *within* the sleepless population. This is emblematised in the contrast and ultimate contestation between Leisha Camden and Jennifer Sharifi. Leisha is the humanised, idealised sleepless, who maintains as primary her relationship with her twin, non-enhanced sister, Alice, and with sleepers. She establishes herself as both avatar for justice for sleepers, as well as for the benign communalist potentialities (as she believes unshakeably) of Yagaism. The notion that sleepers are 'Beggars' (the view of the majority of sleepless) she only half-believes. As the narrative progresses, Leisha finds and trusts not only the essential humanity of 'Beggars', but also, particularly through her bond with Alice, the value of this humanity as 'trade'. Leisha is also blonde, slender, and beautiful; the feminised embodiment of a distinctly American order of values and object of desire. As a pivotal figure, she also signifies a transformed gender order in which 'equality' (albeit ambivalently) reconciles and assimilates conventional feminine beauty ideals with (phallic) social power.

Jennifer Sharifi emerges as both counterpoint and nemesis, not only of Leisha, but of sleeper humanity itself. She is 'dark' – the product of an Arab potentate and a 'Western beauty'. Affectively dehumanised, Jennifer is relentless and utterly without humour. She is a fanatical figure, who founds the defensive and paranoid Sanctuary: a closed, survivalist, terrorist – as well as communalist – community of sleepless. Moreover, Jennifer's fanaticism is constituted as a hybrid of Islamic militancy (she is exactingly devotional in prayer and speaks continually and fanatically of a 'holy war' with sleepers) and an extreme neo-Calvinism that repudiates anything less than absolute in excellence and perpetual motion; whether of body, labour or affective bonds. Both the figuration of Jennifer and of Sanctuary itself make explicit reference to the tensions of the Cold War as well as to its more recent (post-Cold War) variant, which pits Islamic religious fundamentalism against the neoliberal West (while at the same time, demonstrating their reciprocal imbrication).

Sanctuary is a society defined by both paranoia and repudiation. It is, in Kristeva's (1992) terms, melancholic. Kristeva explains melancholia as the affective and psychic effect of a paranoid-schizoid spiral; that is, a spiral in which the Other (and Otherness within the self) is always abjected as monstrous. Sanctuary, like Jennifer, maintains the meritocratic values of Yagaism, paradoxically valorising and yet rejecting its individualism, and forms what might be read as a communalist, if not communist, society that wages a cold (and eventually 'hot') war against Leisha and her few supporters who wish to maintain trade with sleepers:

[W]hat obligation do we have to those so weak they don't have anything to trade with us? We're already going to give more than we get; do we have to do it when we get nothing at all? Do we have to take care of their deformed and handicapped and sick and lazy and shiftless with the products of our work?

(Tony Indivino, who is soon to be murdered – martyred – by sleepers, speaking to Leisha on the eve of the creation of Sanctuary, in *Beggars in Spain*, Kress 1993, 39–40)

The Sanctuary community eventually moves off-planet and, on the basis of their superordinate economic utility – derived explicitly from the trait of sleeplessness – attempt to secede from the United States by means of a threat (and intent) to release an unparalleled biological weapon. Jennifer's perception of the sleepless as, at one and the same time, endangered and exploited but biologically superior and entitled, fuels an escalating paranoia and militancy. Eventually (as Kristeva suggests is the quintessential spiral of the paranoid-schizoid position), the abjection of the monstrous outsider (the sleepers) is turned inward, as Jennifer begins to purge Sanctuary of injured or dissenting sleepless, murdering the former and exiling the latter – including her own 'supersleepless' granddaughter Miranda. It is ultimately Miranda who wrests Sanctuary (and the hostage USA) from the grip of Jennifer's paranoia and radical extremism.

It is interesting to consider the ambivalent figuration of Jennifer. She can be read, on the one hand, as a biologically-flawed sleepless. As suggested in the passage cited above, this is indeed how Leisha perceives her. And Kress ultimately repudiates Jennifer on this basis, as the *Beggars* series maintains its empathetic and idealised construction of Leisha, who remains humane and heroic throughout; and indeed, all the more so when she herself is eventually assassinated by another fanatical figure (this time anti-sleepless) in *Beggars and Choosers*. On the other hand, even as she is cast manifestly as Leisha's alter-ego, Jennifer also latently figures the alternative trajectory of the same trait. In this sense, Jennifer figures as the deeper reality not only of the sleepless characteristic, but of both sleeplessness and genetification as twin (and inextricably linked) biopolitical ethics. Thus, contained within its seductive phantasies, the *Beggars* narrative nonetheless suggests that intersecting class and racialised hierarchies are foundational to and logical consequences of both a genetified and neoliberal social order. In turn, the twin drives of both neoliberalism and genetics – epitomised in the (metaphoric) trait of sleeplessness – set the preconditions for paranoid affectivities and resurgent conditions of cold war. *Beggars in Spain* thus offers a commentary on the underpinning paranoid-schizoid conditions of late modernity, following its dystopian political and conceptual trajectories, but also highlighting its profound seductions.

And We Shall Be Changed

> The body can be *read* as a metaphor for the story… . (Sargisson 1996, 148, original emphasis)

Kress' work provides a salient case in point for the power of science fiction, both as object of and as resource for social theory. In the first instance, the *Beggars* trilogy depicts a constellation of bioethical ruptures and anxieties (and, indeed, rather prophetically anticipates others) in the wake of revolutions in both genetics and the technologies of labouring life. Here ambivalences surrounding issues of de/humanisation, the attrition of humane liveability, and the relationships of political governance, social injustice, and science are pointedly articulated not only as body-political questions, but also, in so doing, as epistemic–*affective* questions. The narrativised form itself allows Kress to infuse feminist theoretical debate on science, culture, and politics with an emotional sensibility. This allows readers to grasp not only the blurred boundaries between dystopian and utopian futures (genetified and neoliberal), but also their phantasmatic purchase.

Through the metaphor of sleeplessness, the works offer a complex critical reflection on the epistemic–affective (knowledge–feeling) underpinnings of neoliberal (as well as foundationalist and reactionary) politics *per se*. *Beggars in Spain* provides a pointed allegory not only of 'sleepless' values as an embodied, technological, and political-economic aspiration, but as emergent from and a basis for a distinctive social order. Kress offers the sleepless body as both immortalising and stabilising against its obverse – the sleeping body as destructive and entropic. The counterintuitive character of these constructions provides the dystopic undercurrent that means that the seductive new order remains fantastical, notwithstanding the realist conventions of plot and characterisation that would seem to place sleeplessness in a realm of futural fact. At the same time, the seductions of sleeplessness are diffused into normative assumptions not only of the plot, but of the wider cultural episteme surrounding the gene: the projective fantasies that genetics can and will produce beautiful, disease-free bodies – leaving only the (rather rhetorical) question of whether it *should* produce such bodies. This fantasy (and the answer to this question) is consolidated in the figure of Leisha – the beautiful, empathetic, and heroic centre of the narrative – who, in her unconditional humanity, humanises the dehumanising tendencies of both sleeplessness and technocratic utilitarianism. Leisha is, interestingly, persuasively – however counter-intuitively – the humane ideal and product of what Kress' novels suggest are a foundationally dehumanising science and a social order intrinsically bent towards the paranoid and totalitarian.

The complex counter-positioning of the themes of humanness–humaneness in the Kress works thus mirrors the ambivalent dystopic tendencies of the late-capitalist work ethic in and of itself, and of its associated imperatives to bodily repudiation and transformation. In both the metaphor of sleep and the elaborated references to recombinant genetics and a transformed (post-fossil fuel) energy base, Kress parses out the ways in which late-capitalist values are simultaneously affective, that is, imbricated in imperatives of feeling, as well as materially embedded in structures of action and modes of exchange. Indeed, the question of 'trade' is offered up simultaneously as a site of social contract, a desired cultural ideal, a crisis of knower as well as of knowledge, and an ethical affective stance. In so doing, the *Beggars* fantasy locates the analytic currents of the trilogy at a 'real world' time of epistemic as well as political and economic transformation and rupture. Kress anticipates a coming 'crisis' of neoliberal capitalism as the ambivalent effect of its seductions. The conjunction of magic and realism in the narrative treatment of technology (genetics and fuel production) paints a disturbing (and compelling) portrait of the ways in which improbabilities, and indeed impossibilities, can become plausible objects (ends and means) of phantasmatic projection and belief. If the neoliberal ideal can only be real*ised* by bodies in perpetual motion and technologies that functionally defy (and eradicate) the essentially entropic character of life itself, then the *Beggars* world both obscures and yet tells us something of the self-deluding character of our attachment to sciences as well as modes of governance that, on closer scrutiny, prove to be based in no more than consoling fantasy.

Finally, it is not incidental that social injustices and the possibilities of their redemption articulate (im)plausibly over bodies. The *Beggars* world invokes a dangerous and powerfully ambivalent bioethic in which the body presages an immanent social world, and is also an artefact — disquieting in each instance — of both utopian and dystopian desire. These are dissonances arising not simply out of spectacular technological or social ruptures, but out of what is (already) normative, and the ways in which what apparently disturbs, also fits. In this way, the *Beggars* allegory sheds some light not simply on what would seem to threaten comfortable lives, but on the comfort that is, and can be, powerfully, normatively, taken in destructive things.

Kress' works do not offer us a simple way out of the bioethical impasse they depict. But in the portrait of that impasse as a vivid and profound tension of feeling as well as knowledge and social praxis, there is perhaps an immanent alternative. This potential may arguably arise as a latent tendency of the science fiction genre itself, with its intersection of allusive speculation and projective identification applied to socio-ethical critique. As science fiction, Kress' works clearly are not simply a thought experiment, but a *thought–feeling* experiment that enables us to apprehend the profound dissonances and dystopic underpinnings of both genetic and neoliberal 'revolutions'. Sleepless values are embedded in both, presaging potentially catastrophic material consequences of which the using up of bodies unto death is only one.

Coda

"If we could identify the gene [for the developmental aging process] and then at young adulthood we could silence the expression of developmental inertia, find an off-switch, when you do that, there is perfect homeostasis and you are biologically immortal," Walker said.

Basically, humans would stop aging past their peak of maturation. Once everything is formed, a human would remain that way for eternity. Of course, death could still come from disease and roller-coasters, but you'd remain at your physical peak until then.

… The parents of Gabby Williams wish researchers would use insights into her condition to focus on helping Alzheimer's patients instead of seeking immortality. As Catholics, they believe in the need for death. But researchers want to take full advantage of the opportunity Gabby has given them. Her mutation is extremely rare and not hereditary, and comes only a few times in a generation.

(Rivlin-Nadler, 18 August 2013, *Gawker*)

This article (one of a number appearing on the same day and also reported on ABC news)[18] refers to genetic research currently being carried out by Dr Richard Walker[19] on 'syndrome x' – a rare condition in which the developmental process is dramatically slowed, such that those affected remain essentially in a state of infancy.[20] Walker's cited comments strikingly echo those of the *Beggars'* fictional Ong, not only in terms of hyperbolic projection, but also in their persuasive normotic tenor. For Gabby Williams (the non-fictional subject of Walker's research), the putative genetic 'off-switch' has not, for example, produced a physical 'peak' but rather its obverse. Yet the dissonance between Williams' condition and the perceived scientific 'opportunity' she represents seems to lend itself to (indeed to normalise), rather than to obviate, extravagant claims. The *phantasy of action* that reconciles that dissonance – that *fills in* its negative space – is the *Beggars* fiction. It is also, manifestly, real.

---

18 Susan Donaldson James (16 August 2013) '8 Year Old Never Ages, Could Reveal "Biological Immortality"'. A documentary segment from *Good Morning America* with the same title is embedded in the article.

19 Dr Walker is based at the University of South Florida College of Medicine.

20 The condition is also associated with additional serious medical conditions. (*Good Morning America*. August 2009. 'The Amazing Girl Who Doesn't Age', ABC).

Chapter 7
Seeking the Jew's Gene:[1]
Science, Spectacle, Redemption[2]

I am not interested in determining the line between "real" and "fabled" aspects of the Jew. This can be done only by ignoring the fact that all aspects of the Jew, whether real or invented, are the locus of difference. (Gilman 1991, p. 2)

Once described as "a sort of British Indiana Jones," University of London scholar Tudor Parfitt, in this interview with NOVA producer David Espar, recounts his fascinating odyssey on behalf of the Lemba. A southern African tribe with tantalizing claims to an ancient Jewish heritage, the Lemba dispatched Parfitt on a journey of discovery that would take him halfway across Africa and into a remote desert valley in southern Arabia, where he stumbled upon what he believes may be the lost city of the Lemba. (NOVA <www.pbs.org/wgbh/nova/israel/parfitt.html>)

Introduction

This chapter examines the collision of spectacle, science and racial-ethnic identifications in the contemporary scientific search for a 'Jewish gene' It aims not so much to distinguish the 'line between "real" and "fabled" aspects of the Jew' (as cited in the passage by Gilman above), but to consider the inextricability of both as composite elements, mutually constituting 'difference' *as* racial-ethnic identification. Thus I am concerned with the specular[3] economies of science

---

1 Sander Gilman's historically framed examination of the *The Jew's Body* provides not only a key theoretical resource for the arguments of this chapter, but a direct referent for its title.

2 A earlier version of this chapter appeared as 'The Search for the Jew's Gene: Science, Spectacle and the Ethnic Other' (2009), *Media Tropes* 2(1): 1–23.

3 I refer here to Luce Irigaray's theory of the 'specular' which offers a re-theorisation of the relationship between gender order and the symbolic. My use of the term here makes an analogy between the phallocentric symbolic (described by Irigaray) and the racial imaginary: both involve articulations of visuality and embodiment, or as Irigaray proposes, between 'light and touch' (Irigaray 1985; see also Zaikin 1999). In turn visuality and touch are imbricated in regimes of gender. In this light Irigaray retheorises visuality (a key preoccupation of psychoanalysis and understood as a superordinate locus of desire, identity and social relations) as a sphere simultaneously of affectivity and of the body.

as well as the knowledge capital of its mediatisation, as they come together, troubled, over the Jew's body. The chapter takes as its case study the 1999 National Geographic (NOVA/PBS) television documentary *To the Ends of the Earth: Search for the Sons of Abraham*,[4] a programme that follows the progress of anthropologist Tudor Parfitt through the Lemba communities of South Africa in a quest to obtain genetic evidence in order to authenticate (or falsify) their claims to Jewish identity.

My discussion will aim to situate the science of what may be termed 'ethnic [or cultural] identity genetics' against a number of historical as well as contemporary resonances. First of these is the use of a (post-)colonial popular imaginary as a lens through which bodies and cultural identities may be taxonomised. Of particular interest in this context (as with the work of Steve Jones) is the revivification of racial science as a legitimate and desired site of human classification. A second arena of resonance concerns the implications arising from a fascination specifically with Jewish bodies in a post-Holocaust world, and with black bodies against a history of colonial eugenic science. Here questions of bodily governance are central, evoking both a complex (and often understood as 'tainted') history of eugenic (sexual–racial) regulation effected by the state and medicine, and the profound forms of abjection and prurience that have historically attended the stigmatised bodies and identities of racial science.[5] Third is the question of representation and the place of spectacle and desire in the sedimentation of scientific ideas – in this instance genetics (and historically, eugenics) – into popular idiom and wider cultural commonsense. As this chapter will argue, all of these themes are articulated on a terrain in which racial–sexual knowledges are deployed through representational as well as scientific economies; and through regimes of desire as well as bodies of knowledge.

This chapter will not undertake an elaborated examination of the scientific validity of the claimed finding of a Jewish gene. While this is an interesting question in its own right, this chapter is concerned with a different (albeit not unrelated) consideration – that is, the question of a 'Jewish gene' as *truth*, as distinct from the question of a 'Jewish gene' as *fact* – a key point to which this chapter will return in the concluding discussion. I begin from the proposition (and will trace the ways in which) the spectacle of science is not simply an epiphenomenal artefact, tacked on to 'real science', but is, rather, part of the

---

4 As a short-hand, I will refer to the documentary by its subtitle *Sons of Abraham* throughout the rest of the chapter.

5 The intersection of race and sexuality can be said to accrue *per se* to the theoretical ideas and historical practices of eugenics (see, for example, Mort 1987; Proctor 1988; Davis 1990; Steinberg 1997) and colonial science (Scheibinger 2004).

epistemic core of scientific cultures and scientific work.[6] Thus, as I have earlier suggested in this book, discussion of a site in which spectacle and scientific work explicitly come together tells us something about *both*.

My discussion begins by introducing the key concepts and media-based analytic strategies as they are taken up in this chapter (some of which have been introduced in earlier chapters). This will include a brief consideration of the documentary genre and the value of social semiotic analysis for an understanding of the question of genes in popular culture. I shall also discuss the particular form of media study I take up here, which brings together feminist traditions of semiotic analysis, an extension of the Foucauldian notion of cultural *episteme* and psychoanalytic concepts of 'gaze' and the role of phantasy. I will then turn to the documentary itself, unpacking the intersecting repertoires of meaning – spectacular, semiotic and narrative – that attach to and constitute the Jew's body and the Jew's gene, as well as the scientific enterprise of identity genetics. Here I shall consider the ways in which the *Sons of Abraham* resonates with and attempts to recast historical imageries and ideas that attached to earlier periods of eugenic sciences, including the abjecting discourses that accrued to nineteenth-century ideas about African bodies and twentieth-century classifications of the Jew's body.

The chapter will conclude by taking up the questions of fantasy (and phantasy) to consider the affective dimensions of both science and its mediatised spectacle. Here I will speculate on the underpinning economics of attachment that may explain both the extraordinarily uncritical reiteration of racial science represented in this particular genetic experiment (and its mediatised representation), as well as the considerable popularity and apparent purchase of racial–sexual identity genetics itself.

Race, Sexual Science and the Social Semiotic Field

Central to this chapter (and indeed to this book) is a methodological question: by what means do we gain purchase on the power relations and seductions of biomedical discourses? What are the processes we aim to theorise in this context to begin to explain the forms bio-power may take, the modes through which the products and processes of biomedical discourse are regulated and reproduced (or challenged); and the particular attractions they seem to embody? As I note above, my discussion of *Sons of Abraham* and the scientific quest for a 'Jewish gene' draws together three themes: the field of visual representation;

---

6 In this sense, my discussion does have bearing on the question of scientific validity as it considers, as we shall see, the problematic underpinning premises for the biological taxonomisation of the search for a Jewish gene.

the question of epistemology – that is, the field of knowledge and the regimes that produce what may be knowable, and the question of affectivity, that is, the feeling structures underpinning regimes of knowledge and power that seduce (or repel) and guide our attachments (and resistances) to particular knowledge forms.

The analytic approach of this chapter draws its impetus from a tradition of feminist semiotics, that is the study of signification[7] informed by a particular focus on the complex articulations of gender, sexuality, race and class (among other key social relations) as they constitute a *signification field*. Such approaches are concerned with repertoires of meaning effected by various means including visual, generic, discursive and narrational. An important strand of this tradition is interested in the nexus of social and semiotic practices, that is, in the relationship between representation and material relations.[8] In this context, the documentary form takes on particular resonance. In her early work, Pratiba Parmar (1987) argued that the documentary is a genre of visual culture whose particular capital is 'truth'. Documentary convention is, in this sense, a conceit not only of real*ism* as a genre style, but paradoxically, a media form that purports to offer *un*mediated facts. As such, it is a genre form that stands at the fulcrum between spectacle and the social. Documentary convention embodies (even as its conventions obscure) the influence of the material conditions and context of its production as well as the consequentialities of its terms of representation.

Documentary, moreover, has a distinctive role in the mediation of scientific work and popular commonsense and at a number of levels. As I suggested in Chapter 4's examination of reportage, media and popular representation is a central avenue through which the largely closed professional circles of scientific ideas become available to and sedimented through the wider culture. At the same time, the way that scientific work is representationally framed suggests something not only about the role and power of media industries, but also about the objects of its representation. In this context, the *Sons of Abraham* (as with other comparable programming) is of particular interest because it represents a moment of overt nexus between scientific work (it documents an empirical research programme in progress) and the work of popular representation (it is a mediatised spectacle). In *Sons of Abraham*, as we shall see, we are presented with an explicit interplay, on the one hand, of a scientific agency taking up the role

---

7 I refer here to Barthes' (1972 edition) understanding of 'signs' (linguistic and visual) as a locus and avenue of cultural meaning-making. Feminist Semiotics has taken this up with particular interest in the gender politics of signification – as both an interpretive field and a site of meaning consumption as well as production.

8 For more extended discussion of social semiotic analysis and its concern with the 'materialities of signification', see Epstein and Steinberg (2007).

of cultural authorship, and on the other, of a framing of scientific work itself with reference to popular ideas.

A linked strand of feminist media studies has been the concern with the question of attachment, that is the affective dimensions of both textual encoding and audience engagement. Theories of spectacle, and in particular, Laura Mulvey's (1988) seminal concept of 'the (male) gaze', highlight the ways in which visual cultures effect meanings through the mediation of feeling, particularly, Mulvey suggests, through a patriarchally inflected mode of desire. In this context, Mulvey theorised visual representation[9] as a site of conflictual subject–object relations, articulated through the voyeuristic pleasures of 'looking' as well as the unconscious identifications of phantasy. Following this, she has suggested that visual representation, organised through a predominating 'male gaze' provides a window on the patriarchal unconscious of the culture that produced it. Another way of understanding this is to suggest that visual representations both constitute and reflect the cultural *episteme* (that is, the *conditions of possibility of knowledge* as well as its consequences). An account of spectacle – that is, the feeling structures imbricated within and through spectacle – thus fills out our understanding of the relationship of the semiotic to the social. As I shall argue, *Sons of Abraham* does not simply present a popularised *account* of a scientific experiment, but in so doing elaborates a realm of desire and phantasy that, together, transforms the advent of a new form of racial–sexual science into both an object of desire as well as an object of plausible knowledge.

A final and related point with respect to media analysis refers to *filling in*, a concept I have cited earlier, and which, along with spectacle, plays a part in forging what might be termed the *affective–epistemic contract* between film and film-viewer. To a significant degree, cinematic signification deals in narrative and semiotic fragments which are then *filled in* by the viewer. This is one way of defining the notion of *popular* in the context of representation: the more recognisable the fragment (the more commonsense it is), the more easily the audience can *fill in* the rest.[10] A documentary which aims to bring the more or

---

9 While Mulvey was specifically concerned to theorise cinematic representation, I would suggest that her work has resonance and application for the theorisation of visual representation in a more general sense.

10 I would suggest that this quality of 'filling in' as a mode of relationship as well as a measure of commonsense accrues to *narrative* form and perhaps also to other forms of linguistic utterance, such as theoretical analysis or performatives. I would include in this formulation forms of visual narrative, and perhaps to musical genres as they are incorporated into visual narrative. This take on the process of affective attachment as well as to plausible indices for a notion of 'popular' does not necessarily work if we consider examples such as music (apart from associations with visual/narrative) or still-framed artistic renderings such as painting.

less unfamiliar – genetic research in this instance – into the realm of popular understanding involves an interplay of the *as-yet-unspelled-out* with *already-familiar* (i.e. cultural repertoires that do not require spelling out). An examination of signifying *tropes* (those encapsulated frames of representation that are suggestive of larger stories or ideas) provides a useful mechanism to gauge not only the terms of congruence forged between arcane and available knowledges, but also the terms of investment. If there are prurient pleasures attached to voyeuristic spectacle, there are also epistemic satisfactions attached to (the quest for) knowing. As we shall see, *Sons of Abraham* is interesting not simply for the ways in which its spectacle might eroticise race, but also for its mutual terms of invitation between the popular and the expert.

Seeking the Sons of Abraham

Scene 1: Land Rover Sequence

A Land Rover comes into shot. It is driven by a tall, casually dressed white man, Tudor Parfitt, who, we already know, is an anthropologist working with a genetic research team at University College London. His attire is Western, with a button-down shirt tucked neatly into jeans. The landscape is a dusty, African wilderness, with parched land and hot sun, the occasional tree and no discernible landmarks. As the Land Rover progresses, the driver glances periodically at a torn scrap of paper which he holds in one hand, muttering audibly under his breath about the difficulties of following this evidently hand drawn map. A sequence of five shots follows: we see the Land Rover going forward out of shot. The scene cuts and we see the Land Rover doubling back, driving again out of shot. The scene cuts to the Land Rover driving left, then doubled back again to the right. Finally it heads back again toward the viewer.

Scene 2: DNA Swab Sequence

The setting is outdoors at night, with a visible moon. The scientist who had been driving the Land Rover in the previously described sequence is now seated at a table, with scientific instruments laid out. He is wearing surgical gloves. There is a group of African men waiting patiently. Each one steps forward and the scientist takes a swab from his cheek. The scientist then poses each man for a photograph. In the midst of this process, the scientist comments jokingly to one man that he must be careful not to mix his "Welsh genes" with their "Lemba genes".

There are a number of observations one might discern from these two scenes. First is that both are obviously staged, and their stagedness is set up around a 'catch' intended to drive particular and recognisable narratives. The Land Rover sequence offers a visual 'play' on a trope of African 'backwardness', a place without proper maps, whose resources remain untapped or wasted by its own inhabitants, requiring the interjection of a Western outsider to trace discernible paths and whose own, imposing and well-heeled presence provides a counterpoint and comment. Against this backdrop, the Land Rover itself seems to evoke both the safari suit and rifle of a previous era, offered here as the late modern accoutrement of colonial adventure and rugged science. Likewise, in the DNA swab sequence, the unlikely use of moonlight to conduct a scientific experiment invokes an air of mystery and adds a certain rough uncivilisedness to the proceedings. In the public swabbing and photographing of African men hovers the shadow of an earlier era of such photography, in which African men and women were posed, holding measuring devices against their skulls, the paraded curiosities of a Victorian racial imaginary.

Both of these scenes appear in the Channel 4 (PBS/NOVA) documentary *To the Ends of the Earth: Search for the Sons of Abraham*.[11] The film is structured around two parallel narrative trajectories. First is the story of a scientific mission led by anthropologist and linguist, Tudor Parfitt,[12] to seek genetic evidence in order to falsify or authenticate the claims of the Lemba community in South Africa to Jewish identity. Specifically, Parfitt and his team were interested in determining whether male members of the Lemba community carried the 'Kohanim gene', that is a purported patrilinial genetic marker, of the Kohen line.[13] In this endeavour, the camera follows Parfitt's journey through South Africa to take DNA samples from various Lemba

---

11 In Britain, this documentary was first screened on Channel 4, 15 March 1999, under the title *To the Ends of the Earth*. It thereafter appeared on multiple occasions for the cable channel National Geographic. It is also available on VHS (NTSC US format only) and can be purchased through NOVA. This chapter is based on transcriptions I myself made from the original British broadcast; another version of the transcript (with a slightly different introductory scene) appears on the NOVA website at <www.pbs. org/wgbh/nova/transcripts/2706israel.html>. It is noted on this website that the US version was screened on PBS on February 22, 2000.

12 Tudor Parfitt is now Emeritus Professor of Modern Jewish Studies at the School of Oriental and African Studies (SOAS), University of London.

13 Kohanim refers to the descendents of Aaron (of the Levite tribe) and, more specifically, those of a 'priestly' lineage and status who served in the Tabernacle. While both males and females can be descendents of Kohanim, the priestly status is patrilinial. Women were not permitted to perform the Kohanim rites and responsibilities in the Temple. (Thus the Kohanim status stands in distinction from those classifications of Jewish heritage and identity that are defined matrilineally).

communities. This scientific journey is both narratively and visually cast through genre conventions of adventure. Here two quintessentially masculine versions of the genre are brought together, emphasising on the one hand the rugged heroism required of (scientific) exploration in a dangerous field, and on the other the unassuming masculinity of the English scientific gentleman, displaced from the otherwise unspectacular mundanity of laboratory life in London.[14] Parfitt's visual distinction from the 'natives' presents as a forceful and continuing visual motif. It is elaborated by his singular whiteness against sequential collectivities of black bodies; a style of casual dress that both concedes to and yet does not 'fit' the rough landscapes over which he travels; his class distinction articulated through an educated British accent; and a style of unassuming and yet entirely taken-for-granted authority – the 'rightness' of this quest is a premise of his presence, rather than an entreaty, or a justification he must submit.[15]

The scientific quest narrative is a paralleled by a second storyline: Parfitt's quest to retrace the Lemba exodus story of their migration from a land called Sena. Thus Parfitt's travels through Africa to gather Lemba DNA are framed and effected through a reverse travelogue of the story of exodus from Sena. This too is an authentication quest, this time to map empirically, a mythology central to Lemba identity. In this context Sena is presented as a metaphor of bodily lineage: both are 'mythologies' of origin and identity; both are framed as unevidenced 'oddities' requiring explanation.

> *V/O (cont)*   Tudor will trace the Lemba train northwards, armed with a genetic sampling kit from the lab. He will try to discover the lost city of Sena. And by taking samples as he travels, he may solve the puzzle of who the Lemba are. And where they come from *(cut to Landrover)*. (*Sons of Abraham*)

---

14 As discussed in Chapter 2, the latter version of this trope is taken up graphically in the (self-)representations of Steve Jones.

15 In a telling scene, midway through the film, Parfitt is 'held up' by one of the Lemba communities he visits as they inform him that they wish to consider whether they are prepared to give him the samples he has come for. Parfitt betrays evident impatience and is heard to mutter in an aside comment that ' … he [the community leader who is taking his time over this] is not quite sober.' The description of the hesitation of the community is described as a 'tribal' sensibility. Because it is presented through Parfitt's point of view, it carries the inevitable connotations attached to the 'primitive' African in the face of the civilised European. The guiding presumption, played out in this scene, is one of obligation on the part of the Lemba; this follows from the logic of a narrative that demands Parfitt's success in getting what he needs. The Lemba are both visually and narratively presented as recalcitrant figures (and figuratives) in direct contrast to Parfitt's embodied (and entitled) knower.

Sena, is described in epic and romantic terms. It is a 'lost' Eden, a mythic seat of the origins of humanity itself, a primordial object of desire. It is also left interestingly unfilled out in the sequence of the film. The particular mythology of Sena is never elaborated, only cited as the film progresses. The vast and mysterious landscapes that appear, at times of Africa as 'dark continent', at others of Arabic lands, are presented as virtually interchangeable. They are only notional citations of land – attached to mystery and to a need to believe. As tropes of religious signification, the indeterminate but insistent imagery of the search for Sena imbues the science with a 'higher' order of desire figured as the stuff of faith. Sena thus frames the putative 'facts' of the Lemba with an aura of larger 'truth', and a projective desire to *believe in* both the Lemba as locus of human meaning itself, and in genetics as a means by which we might grasp it.

At the same time, the notion that Sena is an implausible, untraceable place parallels the implicit (though no less powerful for that) premise of the research: that the Lemba are bodily implausible Jews. In both contexts, it is Parfitt, rather than the Lemba communities themselves, who is vested with the necessary authority to define what counts as evidence and what may be made of it.

One might note also that the notion applied here, that an ethnic/religious mythology must be empirically proven in order for the identity itself to be valid, would invalidate most religious identities. This is precisely the irreconcilable distinction between scientific and religious epistemology. While both constitute 'truth regimes', they are imbricated in radically divergent orders of 'truth', knowledge and materiality. This application *ad absurdum* of a scientific method to what may well be an allegorical account, referencing but not describing a religious heritage, entwines with the racialised economies of the documentary, which are premised on a presumption of invalidity which is much more globally attributed to the Lemba.

The emergent characterisation of evidence here is, itself, striking at a number of levels and emblematised in a scene where Parfitt interviews a Lemba leader about Sena, commenting:

> I was very moved by William. He was clearly very convinced by his story and is genuine in his belief that he's Jewish and his people are of Jewish extraction. And it's interesting he keeps talking about this Sena, this Sena myth, this Sena legend and the story of the Lemba. But it's difficult to see how he can know that. I'm not convinced by this biblical evidence. (Parfitt, in *Sons of Abraham*)

Here we see the slippage of defining agency from 'native' to expert. It is Parfitt who is the definitive 'I' in this formulation, the 'I' who *must be convinced* in order that the claims under discussion may be deemed authentic. The underpinning presumption that the Lemba are *unfit*[16] asserts a motif that will be reiterated throughout *Sons of Abraham*.[17]

Ab-Origine

These underpinning premises of the film with respect to its key figures, its science, and its narrational and visual trajectories are set out graphically in the film's opening sequence:

Opening Scene: Jerusalem (wide angle: wailing wall: roving shots of Jewish worshipers and blowing of shofar)

V/O[18]   History dealt harshly with the people of Israel. Over centuries, they were scattered to the four corners of the earth. Many simply vanished. They became known as the lost tribes. The mystery of what happened to the lost tribes of Israel have haunted the human imagination ever since.

[cue programme title: SEARCH FOR THE JEWISH GENE]

Sintamule, South Africa

V/O   Thousands of miles from Israel, in Southern Africa, live the Lemba people. They believe that their roots go back to ancient Judea.

[shot of unidentified man from Lemba community]

L1   The Lemba are Jews. We are the descendants of Abraham, Isaac, Jacob and the rest of them all. These are our blood relatives.

---

16 The inference of 'unfitness' seems to carry two intersecting constructions of the Lemba. One is that they are *unable* to authenticate their own identifications – both by reason of lack of necessary expertise and knowledges – and another, related construction, that they are *unwilling* to do so. This latter is implicit in the ceding of self-definitional authority to Parfitt as well as the oft-noted perplexity that the Lemba lack the curiosity to pinpoint geographically the location of Sena. This of course begs the question as to why a biblical mythology must be empirically validated in order that the identities associated with it be deemed to have authenticity.

17 Again, even as it establishes a 'higher order' of desire, the conceit that Sena itself must be empirically located is, of course also a rather odd and absurd one given that religious mythologies by definition constitute a faith based episteme which is not reconcilable with scientific understandings or methods of knowledge.

18 V/O = voice over; T = Tudor Parfitt; L = member of Lemba Community.

V/O   Their lives for the most part are traditionally African. <u>But</u> the Lemba have customs which <u>could</u> connect them with the Jews. They follow the strict dietary rules that were laid down by Moses. They eat only kosher food. All blood has to be drained from animals before they're fit for consumption. *(original emphasis)*

Soweto, Johannesburg [shot of Tudor Parfitt, tall, white with educated British accent welcomed by Lemba]

T   Good afternoon.

V/O   Tudor Parfitt is an anthropologist and linguist who's investigated *claims* to Jewish origins all over the world. Many diverse peoples from Jerusalem to Japan *claim* to be descended from the ancient Israelites. *It may be a need to belong rather than reality*, but after many years spent recording the Lemba's oral history and customs, Tudor has begun to share *their extraordinary belief. (emphasis mine)*

[cue: Tudor Parfitt centre of group of Lemba]

T   You're called the black Jews. Do you <u>really</u> think you are black Jews? *(original emphasis)*

[cue: a second unidentified Lemba man]

L2   I believe that I am a black Jew. Because we don't eat pork and the Jews also doesn't eat pork. There was these marriages who don't [unclear] marry of different nationalities. It must also be of a Lemba tribe. And whenever we slaughter we wash our hands with care as well as the utensils. To eat kosher food, they clean all the utensils, they clean their hands and so on. And that means everything is clean.

[cue: community gathering in large square addressed by Lemba leader. Background voice counterpoint to voice-over, of translator — not in shot — who explains leader's speech in English]

V/O   In modern day Soweto, the Lemba are determined to preserve their Jewish identity. Their leaders remind them that Africa was not their original homeland. They tell of a journey across the sea from a place somewhere in the North called Sena.

Translator   We started from Sena. Then we get to Sena 2, Sena 3 …

V/O   Today, no Lemba knows exactly where Sena is or was. *In spite of their passionate belief* that they are Jews, **Tudor** has had no means to prove or disprove their claims ... [pause] ... **until now.**[19]

In this opening sequence, we see the conjoining, both through narrative structure and generic (linguistic and visual) repertoire, of imperial and scientific adventure. What is set out is a scientist on a mission to a 'dark continent' where he meets reluctant, childlike natives who do not understand their land or its natural resources and so consequently their lives are marginal and poor. It is for the scientist, who embodies the contrasting nature of a civilised, resourceful and knowing West, to make use of these resources; to mine what is subterranean; to define what is amorphous; and to develop what has been left untended in the desert. In this context it is the scientific narrative that transforms the journey to find Sena into an appropriative and territorialising modus, through which the Lemba become objects, rather than subjects, of their own history. At the same time, these two journeys, one to mine DNA and one to retrace the road to a mythical place of origin, are offered as redemptive, transvalued quests – on this occasion ostensibly to restore rather than to plunder; to offer goods rather than to appropriate; to authenticate rather than to devalue; to right the wrongs of an earlier imperialism.

A constellation of unsustainable premises accrue to this understanding of bodies and to genetic knowledges of bodies. By both scientific and filmic convention we are asked to take a number of things on faith. For example, embedded in the quest for a 'Kohanim gene' is a prior assumption that cultural identity carries (and can be reducible to) a biological marker that can be, and this is the second assumption, scientifically pinpointed and accurately traced. It is a presumption rather than an argument – of both the science and of the documentary tracing it – that genetics can tell us something about Jewish identity; who is a Jew, how it is passed along and what it means. The positing of an indexical Jewish gene against which claimants to the identity can be measured rests on two further assumptions. One is that 'Jew' and 'Kohanim' consitute empirically coherent categorisations[20] – an assumption that rests on yet another presumption – that of 'unbrokenness' of reproductive relations and kinship from Jew to Jew, from Kohen father to Kohen son. A related presumption is that the Lemba, as a people, constitute an 'undiluted' culture, back to their 'primitive' roots. Terminologies of 'mystery' and 'lost tribes' juxtaposed with

---

19   This excerpt is taken from the British version of *Sons of Abraham*, screened on Channel 4, 15 March 1999, as *To the Ends of the Earth*. Italics indicate original emphasis, bold print indicates my emphasis.

20   In actual fact, Jewish history is a history of conversion into and out of Jewish identities.

haunting, stark landscapes and references to human imagination itself evoke an epic antiquity and the profound power of an encounter with the slender and inchoate remnants of humanity itself. It is perhaps worth adding that the idea that exact paternity, particularly back into antiquity, can either be *known* or can be *assumed to have followed* for the purpose of the research seems an extraordinary one, given what we know about the realities of human sexual relationships.

Anthropomorphism and Ethnic Capital: The Jew's Body and the African 'Other'

The notion of a 'throwback', in-bred culture was a characteristic object of early anthropological fascination, as well as of early genetics, and it was not incidental that such researches focused on racialised 'oddities' of under- (and sometimes *uber-*) class.[21] Trinh Minh-ha (1982) has suggested that the anthropological gaze is characterised by overtones of anthropomorphism (that is a gaze on the primitive as not-quite-human), hence its association with the zoologically focused tropes of Natural History. Here the Lemba are figured at one and the same time as the repositors of human origins and as quintessential outsiders – objects in need of explanation, rather than subjects who may (and have standing to) explain themselves, without having the *necessity* of doing so. Moreover the African context for such investigation is as familiar, indeed is a veritable cliché, as is the instrumentation (swab, camera, latex glove, rough-hewn tables by moonlight) of (post)colonial cultural taxonomy and its imbrication in a eugenic racial imaginary.

The Jew as Mongrel Body

In his 1991 study, *The Jew's Body*, Sander Gilman examines the highly-charged characterisations of Jewish identity in the history of racial science. This history is in part located in the nineteenth century, where the Jew's body emerged as an ambivalent figuration against imperative investments in the notion of definitive racial lines between white and black peoples. As a people whose perceived insistence on their difference jarred with the assimilative logic of the emergent modern state, Jews became figures of debased whiteness, and the Jew's body, the object of prurient fascination. It was in this context that the Jew was cast as 'swarthy', defined as a 'mongrel' race – the depraved product of interbreeding

---

21 For discussion of the fascination of early genetics through the simultaneously classed and sexualised fetishisations of Victorian eugenics, see, for example, Mort (1987).

of white and its reviled Other, African.[22] Gilman cites the example of the 'Hottentot' who were seen to emblematise the animality, danger and intrinsic depravity attributed to Africans as 'ugly' race (p. 173). As 'hybridised blacks', Jews too were understood as members of the 'ugly races'. These prurient characteristions in turn formed ideological centres, the latter rationalising, among other things, the slave trade, and the former underpinning the eugenics movements of the early twentieth century – which culminated in the genocidal catastrophe of the Holocaust.[23] Against this background, the Lemba people, as Africans and Jews both, stand as ethnic Others at the interstices of two intersecting racial taxonomic traditions. Their presence, both as objects of genetic-ethnic experiment and as spectacle, cannot but carry the resonances of such earlier associations.

The Lemba Body

Tudor Parfitt is an anthropologist and linguist who's investigated claims to Jewish origins all over the world. Many diverse peoples from Jerusalem to Japan claim to be descended from the ancient Israelites. It may be a need to belong rather than reality, but after many years spent recording the Lemba's oral history and customs, Tudor has begun to share their extraordinary belief.

(v/o narration, *Sons of Abraham*)

*You're called the black Jews. Do you **really** think you are black Jews?*

(Parfitt, *Sons of Abraham*, original emphasis)

In the logic of the documentary (and arguably, of the science it tracks), the Lemba are *axiomatically* implausible Jews. Proceeding from this premise is an extensive language of disqualification that disinvests the Lemba as claiming subjects and locates trustworthy explanation in the educated authority of white European science. As emblematised in the passages above, Lemba self-definition as a Jewish community is cast with the language of suspicion: it is a 'claim', an 'extraordinary belief'; they 'have customs which *could* connect them with the Jews'; 'do [they] *really* think [they] are black Jews?' Implicit in this language is a construction of a credulous and abject people: the comment that 'it may be a need

---

22 The sexualised voyeurism embedded in this discourse is graphically evident in the example of Sarah Bartmann (the 'Hottentot Venus') whose prurient display was emblematic of the repudiative imaginary of Victorian racial ideology, articulated through a nexus of science and spectacle. See Gilman (1985); Holmes (2006).

23 For further discussion of the racial-scientific underpinnings of twentieth-century eugenics and the eugenic and genocidal practices of National Socialism, see also Lifton (1986); Proctor (1988).

to belong rather than reality' is suggestive of a self-denying (but understandable) plea from the marginal 'low' for distinction above their station.

At the empirical centre of this representational economy are bodies (and identifications) that do not 'fit'. The Lemba are expressly posited as implausible Jews *because* they are black. This premise is linguistically grounded through the use of 'black' as a qualifier for 'Jew'. Moreover the 'Africanness' of the Lemba is, throughout the documentary, visually and discursively staged in negatory terms: the Lemba people 'perform Jew',[24] but the performance is somehow 'not right'; their origin story is 'unfamiliar' and 'odd'; their practices are not strictly 'orthodox' (by European Jewish standards) and therefore, by implication, must be *unorthodox*. In this way, and without ever having to say so directly, (European) whiteness is implicitly cast as an indexical characteristic of the 'real' Jew.[25] This is of course an odd construction, given that Jewish identities cross conventional racial boundaries.[26] It is not in any way extraordinary to find Jews across 'racial' typologies. What is particularly notable in this context is the citing of ' Jewish' as an *aspirational* identity and one *in contrast to* 'African' and 'black'.

The Jew as Aspirational Body: Racial Transvaluation and a Reparative Science

From the imputed standpoint of the Lemba, 'Jewish' is an object of desire, a sought status, where authentication, against the presumption of doubt, can only have an *elevating* effect.[27] This suggests one way of interpreting the cited 'need to belong' attributed to the Lemba — that is, as an illegitimate claim to the distinctions of whiteness, but taken 'sideways', by means of a white identification that they seem not to understand is 'tainted'. The Lemba are not figured here as cynical. Rather they seem to (innocently) collude in the notion

---

24 In a characteristic scene, for example, a Lemba man appears in full religious regalia, but outside a context where such attire would normally be worn. This projects an air of 'unfittingness' to his descriptions of their Jewish religious and cultural practices.

25 The complexities of 'Jewish' as multiple identities and contestatory cultural definitions, heritages and practices — indeed as a cultural *phenomenon*, are entirely residualised by this emphasis and elision with 'Jewish' as singular (and unifying) biological marker.

26 Jews can be found across most if not all conventional 'racial' as well as national classifications.

27 It can be argued that this aspirational association is further sedimented in the person of Parfitt, who is already figured as aspirational figure in contrast to the communities through which he travels. In this context, he appears also to symbolically *stand in for* the specific distinction sought — an aspirational whiteness, here conferred by the mark of the Jew.

that 'Jewish' is 'white' (until proven otherwise) and in the implicit contrast of value made between Jewish and African identity. Their identification is figured as *understandable* wishful thinking. This implicitly reconfirms the 'rightness' of racial attributions in (and as) conventional hierarchies of distinction between white and black, even as it leaves the Jew in a familiar, tainted netherworld between both.

Thus the construction of the Jew as an aspirational identification reflects both *transvalued*[28] as well as more conventionally derogatory associations. Jew as *desired* rather than *reviled* status presents an obvious contrast to the racialised history of the Jew, but it is premised on a standard cliché of African as 'backward Other'. Indeed, *Sons of Abraham* is strikingly silent on the catastrophic historical consequences of racial science, for both Jews and Africans. And it is difficult to know what to make of this. It is, for example, explicit in the narrative logic that this is a quest *to right a wrong*. Two possible wrongs are implied. One is the possibility of a people who are claiming a 'wrong' identity. The other, the *preferred wrong*, given the moral trajectory of the film, is that here, we have a people that is potentially *wronged by* the invalidation of their ('odd' – but real) identity. Thus, the scientific quest to provide 'right' answers about the Lemba cannot but carry an aura of a deeper purpose, a mission to right larger wrongs – the wrongs of racism (and racial science) themselves.

As viewers, we are powerfully positioned *with the grain* of redemptive desire on offer. These are mutually reinforcing prospects of both vindication for the Lemba (to be both transvalued and, by genetic means, proven 'right') and, in so doing, of the transvalued recuperation of genetic science from its tainted past. In this context, the desires of Tudor Parfitt and of the Lemba are positioned in a complementary affective terrain; both are subjects of identification as well as mirroring loci for satisfactory narrative closure. We *want* the Lemba to be 'real' Jews.

V/O   By now, Neil Bradman's geneticists have completed their work on the African samples and they've come up with a **stunning result**. The Lemba Y chromosomes do show a number of links with middle eastern peoples, including Jews. But one of the Lemba clans, Professor Matibha's Buba priests possesses that unique genetic signature, the Kohanim gene. Inherited from the Israelite priests who served in the temple 3000 years ago.

---

28 I refer here to Gilman's (1991) discussion of 'transvaluative racial profiling', as a discursive re-valuation, in positive, aspirational terms, of formally stigmatised racial identifications. Gilman cites the politics of 'black is beautiful' in the 1960s as an example of a transvalued racial discourse. He argues that racial discourse has been historically subject to attributions of both positive and negative qualities.

Modern science has given the Lemba the means to **prove** that their ancestors were indeed among the ancient people of the bible. The Lemba **are** decendents of Abraham. Their genetic trail definitely leads out of Africa back to Israel.

(*Sons of Abraham*, emphasis added)

And this satisfying dénouement is finally, predictably, and indeed by necessity offered. The Lemba, we are told, *are*, after all, 'real' Jews, and indeed, not only that, but Jews of high order – they carry the *Kohen gene*. The closure offered here, however, presents an uneasy conclusion. For a science that can definitively authenticate a Jew's body cannot be so uplifting a prospect. And the reconciling of black and Jew, is not so novel as that. History is not so simply displaced by silence or wishful optimism. Moreover, even as the Lemba are ostensibly vindicated, their marginality to the question of their own self-identification as well as their imputed oddity are expressly reinstated. It is Parfitt and 'Bradman's geneticists' who bear the means of 'truth' and the standing to present it, knowingly, insinuatingly, as a *stunning* twist: *truth*, so goes the cliché, *being stranger than fiction*. This conceit (an homage to pot-boiler and mystery genre alike) would seem to disclaim what is perhaps less obvious, that a genetic invalidation of the Lemba within the narrative logics of this documentary and this science, would have been virtually inconceivable. Such a counter-finding could only have *negated* the authoritative as well as ethical standing of the science (and the scientists); could only have reinstated rather than redeemed the colonial–racial referents that provide the imaginative foundations of the enterprise.

Phantasm and Desire

I would like to conclude this chapter by raising two linked questions, one concerning the terms of congruence between representation and science as they are exemplified in the *Sons of Abraham*; and the second turning to perhaps the more interesting question of desire and the seductions that seem to accrue to the search for a Jewish gene. Here I would like to offer some speculative thoughts on the underpinning fantasies (fanciful images that reflect more or less conscious wishes) and phantasies (the unconscious arena of desire and inchoate yearnings) that seem to be attached both to the science and the spectacle of a search for the *Sons of Abraham*.

What is the significance of a post(colonial) imaginary as a lens for a contemporary taxonomy of culturally-embodied identity? I would suggest that the juxtaposition of this specific scientific experiment and the conventions of its representation are not incidental. The colonial adventure narrative arose as part of the cultural apparatus of imperial expansion and framed 'race' as a eugenic as well as colonial science. This in turn presaged the intersecting histories of

Jewish and African people as bodies of voyeuristic fascination and as degraded identities. In this context both racial science and the material practices accruing to it dovetailed with a particular popular imaginary: the racial *episteme* thus forged the conditions of possibility of knowledge about bodies, cultures and identity – producing the subjects and objects of (post)colonial modernity. In this endeavour, science and spectacle were not only contiguous, but mutually inextricable. The contemporary quest for a 'Jewish gene' would seem embedded in this logic, notwithstanding (or indeed one might suggest, because of) attempts to frame it as a redemptive (or perhaps simply 'innocent') quest. Many questions arise: is it possible to focus a taxonomic gaze on Jewish and African bodies without invoking the history of scientific racism? What responsibilities does contemporary science have to those histories? Does a genetic understanding of cultural identity *intrinsically* tend toward objectification and voyeurism? Was *Sons of Abraham* simply a 'bad' representation of a 'good' science?

A second question arising in this context concerns the specific processes of visual spectacle and the terms through which science sediments through the wider culture. Visual media and the processes of spectacularisation (including through non-visual forms such as language and narrative) are not simply epiphenomenal events tacked on to science after the fact. Rather they reflect and forge the affective economies that both drive and accrue from scientific endeavours. In other words, spectacle has an *epistemic* character: it concerns the relationship of feeling to knowledge (the conditions of what can be know*able*). And *epistemology* has a 'feeling structure'. In this respect, *Sons of Abraham* provides an edifying example of the ways in which racial sensibilities are infused with relations of desire – both voyeuristic (eroticised through the functions of 'the gaze') and epistemic (the satisfactions of knowing, linked to but not reducible to the erotic). The spectacular framing of the quest for the Jewish gene thus reflects a continued currency attached to racial taxonomy: the Jewish gene is an object of desire through its reiteration of (as well as through its apparent departure from) antecedent characterisations of the Jewish and African bodies, themselves articulated through regimes of knowledge attached to desire. That Jewish genes are ascribed capital in this context speaks of race as an arena of continued capital; its apparent transvaluation in this instance seems to present as a minor (and as we saw ambivalent at best) recuperation in an overriding regime of human value ascription, historically given to inhumanities but also so powerfully persuasive, so given to noble expectation. The rehabilitation of racial science offered here is in part possible because of the ways in which, by such means, it can seem to encompass its opposite tendencies (it can be used 'for good').

The visual tropes of a magnificent and primordial African landscape, the epic languages of antiquity and of journeys across seas and vast unmappable distances, and of the foundations of human history itself, powerfully reference what is understood as an inchoate human desire, a projective phantasy of origin in which one finds oneself at the seat of humanness itself. This is constituted at one and the same time as an epistemic as well as fetishistic desire – one that intermingles a drive to know with an eroticised drive to visual delectation. The anthropomorphism of land, and of Sena as a mythology of place, thus cements an anthropomorphism of body and identity. Even as the Lemba emerge as figures of a 'throwback' gaze, they, like Sena, become by this means, imbricated in a mode of desire that promises to bridge the evolutionary distance between 'them' and the notional 'us', the preferred subjects of the filmic gaze, who are not Lemba, but who find in 'their' redemptive explanation, something elevating of 'our' own.

The science of the Jewish gene, as well as the terms of its representation, tap into (and arise from) deeply familiar tropes of racial difference as hierarchy. They play out on a field of the *already-known* with respect to which bodies require explanation and which do not. The reinforcement of the already familiar speaks directly to both the epistemic pleasures of *filling in*; by this means, one, as viewer, can recognise oneself as knowledgeable subject. In the *Sons of Abraham*, the specific terms of *filling in* take place in at least two ways, both of which imbricate knowing and feeling. First are the visual tropes and conventions of anthropological voyeurism that direct our subject–object identifications. There is a notional 'we' here: those who are not Lemba, but who find them a compelling curiosity. This 'we' is confronted with figures who are at once pathetic and *sym*pathetic. Perhaps most importantly, the visual economies of *Sons of Abraham* invoke, without the necessity of spelling them out, a racial commonsense that, renders comprehensible both the desires attributed to the Lemba as well as those of Parfitt and his team.

As an article of faith, the DNA testing carried out here is cast as definitive and proven rather than as an experimental and inexact technology. It is only the Lemba who seem to occupy the arena of 'experimental'; as objects of plausible doubt, their ambiguities are seen to demand intrinsically the intervention of an authoritative investigation. And yet *is* cultural genetic profiling a proven 'fact'? Can something as complex and diverse as cultural identity be 'captured' by a gene, either literally or ideologically? For that matter, to what degree is a 'gene' by any definition as 'proven' fact, given contested views within the scientific community about what genes are. The focus on the Lemba as scientific conundrum displaces the questions of the gene and of genetic testing as matters of discussion, doubt, and contestation. At the very least, DNA testing of this community, whether for purposes of falsification or authentication, constitutes a social experiment, and an unacknowledged one at that. The presumptions of benignity, partly emergent from the narratively predictable outcome (the Lemba *will* be proven to be Jews, and not only Jews, but *Kohanim*) sidesteps the question of just what kind of impact might accrue to the casting of a people's history and culture to the realms of doubt. What would it mean if the Lemba had not been found to have 'Kohen' genes?

Embedded in this spectacular economy is thus a second order of *filling in* – what might be termed the 'economies of expectation' that emerge from the parallel plot trajectories of the narratives in play. The narrative conventions of colonial adventure and of scientific quest turn on linked dramas demanding the facing of danger and the solving of mystery – both inflected by a higher purpose (the seeking of the grail). In this context, the finding of a Jewish gene *against-all-odds* becomes both a foregone conclusion and a necessary dénouement. As viewers, we are positioned with a tide of expectation, in this instance, the desire to find what is being sought and to believe *a priori* in its capital as *truth*. It is notable that nowhere in *Sons of Abraham* is the science of the Jewish gene substantively explained.[29] Instead, the gene as *fact* and the scientific interpretation of it as *certainty* is presented as a logical inference of its status as *true*.[30] And that 'truth' is an admixture of its *plausibility* predicated upon a powerful investment of belief.

Sons of Abraham provides, in this sense, a window on the profoundly phantasmic character of race itself. The Jew's gene (and both Jewish and African bodies) oscillate, deeply troubled, between the experiential materialities of racism and the phantasmic projections of place and standing, the profound affectivities – both of desire and repudiation – that accrue to race and ethnic identifications. We are never told *why* it is important to know if the Lemba are 'really' Jewish because the imperatives of a racialised field are already understood, and already so profoundly embedded in the wider episteme (both scientific and popular). The answer is obvious. The question follows.

The claim of finding a Jewish gene also speaks to the fantasy of predictive rationality and the desire to believe in the intrinsic 'good' of that rationality. Investments in the elegant efficiency of science, in its precision and predictive prowess, are quintessential modern values and deeply felt terms of attachment to science as avatar of what counts as progress. The modern scientific subject arising from this ideal bespeaks a *normotic* phantasy[31] – that is, a phantasy that would deny the possibility of the uncaptured and unknowable, and that, in so doing denies the unknowableness that is characteristic life itself. The scientist as normotic subject is one who *knows* (or can know) absolutely; one whose objective gaze guarantees that he will be the author of his own being. In this

---

29 The more elaborated explanations provided in literary and scientific contexts recapitulate the core presumptions staged in the documentary (see, for example, Parfitt 2002). It is arguable whether more detail on method provides empirical basis for those presumptions.

30 Boler (2006) has analysed this distinction as it arose in the satiric political comedy of Stephen Colbert in his coining of the word 'truthiness' to describe the quality of being 'truthy' as distinct from 'facty'.

31 With thanks to Peter Redman for extended discussion with me on this point.

sense, the project of genetics itself seems imbued with normotic promise, a bulwark against the uncertainties of place that attach to day-to-day lives. It is the promise also that genetics has something to say about who we are, has some pre-eminent capital to add to human distinction.

The 'Jewish gene' appeals to investments in deep notions of identity, to phantasies of transcendence and of belonging, to desires to locate ourselves by way of origin stories that tell us our lives are meaningful, connected to others and part of historical currents bigger than ourselves.[32] There is also the collateral and manifest fantasy that we can know the boundaries of identity because they are marked (as markers for better things) within our own bodies. The notion that what was formerly reviled can be redeemed is yet another powerful notion. In this context, the specific transvaluation of the Jew offers a metaphor of the possibility of transcendent personal redemption, perhaps one of the most powerful iconic aspirations of the Judeo-Christian (and Western) imaginary. Likewise, the possibility of a redemptive science, particularly on this terrain, speaks not only to familiar post-Holocaust and postcolonial discourses of reparation, but in so doing, to the imagined possibility of finding again an ethical life, out of the ashes of human atrocity.

---

32 For poignant further discussion of this question of investedness in identities and identity politics, both as a remedy for social exclusions and as a site for belonging, see Rothman (1998).

Chapter 8
Between Biology and Culture – What is 'Real' About Genes?

Introduction

In its appraisal of the cultural apotheosis of the gene, this book has tracked a number of themes. Chief among these has concerned the articulation of science and culture in and through relations of spectacle and the intermediation of social, political-economic and popular discursive formations. How does science filter into the wider culture; to language, to practice, to the wider imaginary? How does it come to attach to cultural anxieties, to projections of hope or fear, attachment or revulsion? Why do some sciences capture the popular imagination, engage the passions, summon the wider spectrum of investment while others remain obscure. In pursuing such questions, this book has challenged the still hegemonic view that science is an enterprise that is (or should be) separate from the undertakings of culture. Instead, it has argued an alternative understanding of science as embedded in culture; both shaping and shaped by the values and ideas of its time; both mobilised and constrained by the historical–locational conditions of its practice and its articulation with other socio-cultural industries including politics, commerce, and communications. In this final chapter, I would like to consider these themes in the service of yet another question: how do the relations of plausibility and possibility of knowledge constitute what is 'real'?

As I was working on an earlier version of this chapter, I came across a course synopsis, developed by a colleague, which was to be offered as part of a degree in communications. She called the course 'Reading Our World' and her description struck me:

> [This course] examines the links between language and contemporary social and political issues and emphasizes the close relationship between critical thinking, critical reading and critical writing.[1]

The wording has an interesting resonance for the cultural field of genes. Reading emerged and has remained as one of the most prominent metaphors

---

1 With thanks to colleagues at UOIT who shared with me the early planning and materials for this course.

for genes and genetics across both popular and scientific contexts. Indeed, Steve Jones' lectures and book *The Language of the Genes* provided an early and indicative launching point for genetics' most consistent popularising projection. Earlier I suggested that 'reading genes' constitutes, at one and the same time, a persuasive democratising metaphor and yet a pointed obfuscation of the technologically-based and specialised practices and knowledges that constitute the actual work and field of genetics. These are practices that are themselves, in turn, part of a wider spectrum of specialised expertise and emergent 'revolutions' – in informatics, in pharmaco-medical, agricultural, genealogic, criminological and jurisprudential practices that have filtered into everyday narratives and vernaculars and that have in so many respects reconstituted our identities and communal relations, our understandings of health and illness and our relationships to history as well as futurity. Yet 'reading' also provides a useful analogy for the project of this book. Indeed, it foregrounds this chapter's central question of what is 'real' about genes. To answer that, we might first ask, what does it mean to read our world through genes? And by corollary, what is the relationship between reading and authoring?

We live in a world that is now pervaded with genetic languages and practices. Genetics spans medical, scientific, agricultural and pharmaceutical applications; it is as much a part of the food imaginary, for example, as it is of the food chain. Genetic discourse is now fully embedded in our languages of kinship, of culture, of personal identity, of humanity, and of animality. Genes are part of common parlance – they are referents of everyday life. They appear as side and central motifs across genres: in cartoons, in comedy, in school curricula, in work, in health, in the definition of what counts as family and cultural origin, in defining and solving crime and in the law. Genetics, these days, is asserted as much as a tool of history as it is of medicine. Genes are also sites of attachment. People, institutions and modes of governance are vested in genetic explanations. Genes are objects of powerful and deeply felt projective fantasies, aspirations and anxieties.

In what follows, I will return to the cases discussed so far in this book to consider this cultural sedimentation of genetic discourse from yet another vantage point – specifically, the articulations of the gene between biological and cultural imaginaries. In so doing, I aim to consolidate this book's multivalent reflection not only on the epistemic, but also on the affective, spectacular and phantasmic dimensions of genes and genetics, and to explore the ways in which the gene has emerged at the interstices of science and culture and as a product of science *as* culture. The central argument of this chapter, in response to its perhaps somewhat provocative title, is *not* so much that genes 'aren't real' but rather that the reality of genes would seem to owe less to their factuality than to their persuasive power as objects of desire and their consequent plausibility as objects of knowledge.

If it Looked Like What it Was

> Meeting a friend in a corridor, the philosopher Ludwig Wittgenstein said, "Tell me, why do people always say it was natural for men to assume that the sun went round the earth rather than that the earth was rotating?" His friend said, "Well, obviously, because it just looks like the sun is going round the earth." To which the philosopher replied "well, what would it have looked like if it had looked as if the earth was rotating?" (Stoppard, 1986)

This oft-quoted epigram from Tom Stoppard's 1986 play, *Jumpers*, is interesting for a number of reasons. Firstly, it calls into question what it means to see. It demonstrates that facts do not necessarily constitute or correlate with what is real. If they did, we would see the sun and earth and their relationship would 'look like' what it empirically is. Reality is posited instead as a filter, as a *field of intelligibility* that reciprocally imbricates what is *material* (our senses, our experiences, our mutual publics composed of laws, institutions and place) with what is *understood* – as much a projection as it is a consequence of fact. Thus, what is real is both consequential *to* and a consequence *of* what is seen, and not only that, but what is *sought*.

We might ask then, what are the facts of genes? What does a cultural standpoint, if this can be distinguished from a biological standpoint, allow us to understand about the facts of genes? What are the conditions of genes as objects of knowledge (or doubt)? What are the consequential effects of genetics as a field of practice, a body of knowledge? What is the reality of genes – as bodied phenomena; as loci of capital; as objects of governance or transference or faith; as articles of law or commerce or war; as ideas? What are the cultures of the gene? What, moreover, does the 'language of genes', suggest about science, not only as a knowledge enterprise, but as a culture industry?

The Language of Genes

> Genes are likely to influence the occurrence of criminal behaviour in a probabilistic manner by contributing to individual dispositions that make a given individual more or less likely to behave in a criminal manner. (Lyons, *Genetics of Crime and Anti-Social Behaviour*, 1996)

> Tay-Sachs disease is a fatal genetic disorder, most commonly occurring in children, that results in progressive destruction of the nervous system. While anyone can be a carrier of Tay-Sachs, the incidence of the disease is significantly higher among people of eastern European (Ashkenazi) Jewish descent. (National Human Genome Research Institute <www.genome.gov>)

153

The setting is outdoors at night, with a visible moon. A Caucasian London scientist, dressed smartly in casual western attire, is seated at a table. There are scientific instruments laid out and he is wearing surgical gloves. There is a group of African men waiting patiently. They are members of the Lemba people, a tribe that claims Jewish identity. Each one steps forward and the scientist takes a swab from his cheek. The scientist then poses each man for a photograph. In the midst of this process, the scientist comments jokingly that he must be careful not to mix his Welsh genes with their Lemba genes. (*To the Ends of the Earth: The Search for the Sons of Abraham*, 2000, NOVA)

Angelina Jolie's Genetic Self-Ownership Is the Future of Medicine. (reason. com, 15 May 2013).

On a remote jungle island, genetic engineers have created a dinosaur game park. (Crichton, 1991. *Jurassic Park* blurb)

What observations might we make from this juxtaposition of instances? First is their suggestion of the cultural pervasiveness of genetic discourse. Genes and genetics appear in contexts that range across diagnostic medicine, laboratory science, television documentary, celebrity and popular fiction. They cross over questions of kinship, of governance, of labour and political economy, of ethics and of desire. They are characterised in languages of action. They are invoked in terms of technical specification. They are objects of fiction. A second observation might be that the 'gene' emerges as a slippery[2] category, referring to very different phenomena and orders of knowledge. The gene is at once a unit of heredity; a marker of identity; a regulator of cellular processes and protein production; a harbinger of (potential) disease; a predictor of behaviour or orientation or character or progress; a call to action. Third, we might note that there is in all of these examples, a dynamic interplay of biological and cultural referents.

Crime

Michael Lyons, quoted above, was at the time he published his piece, a scientific researcher with the Harvard Institute of Psychiatric Epidemiology and Genetics. The quote is taken from his contribution to the 1996 CIBA Symposium on the *Genetics of Criminal and Antisocial Behaviour*. The symposium, we will recall from Chapter 5, considered a range of topics, ranging over the genetics of mouse aggression, the genetic significance of criminality (as

---

2 In addition to its denotative meaning, I use 'slippery' here also to invoke slippage in the 'leaky', body-reflexive sense described by Shildrick (1997).

derived from and equated with conviction records), 'anti-social behaviour' among twins and adoptees and the genetics of warfare and violence. My earlier discussion of the symposium raised a number of points that are relevant for the questions at the heart of this chapter. First is that the symposium's claims concerning the genetics of criminal and anti-social behaviour were premised on categories that are, at their foundation, cultural. The meanings of 'crime' and 'antisocial' are historically and locationally diverse, radically contested and lacking in the kinds of categorical continuity that would support an inference of biological cause or equivalence. Second, many of the studies relied on highly questionable assumptions about the coherence and reliability of the social field, drawing on conviction statistics as their baseline, following a presumption that convictions unproblematically align with or reflect 'crime', and eliding different orders of 'crime' from markedly different, and differently socially positioned, groups of people. The disproportionate alignment of conviction statistics with social inequalities were also taken for granted. The indexical frame did not, for example, encompass misconduct in banking and finance or transgressions committed by elected officials – which do not tend to translate into convictions. Also elided were the symposium's chief indexical referents – the categories of 'crime', 'anti-social behaviour', 'aggression' and 'violence' were asserted interchangeably. At the same time, none of the participants posited the reductive 'hard determinist' proposition of a single gene for criminality. Rather, they set out the 'soft determinist' proposition that a plurality of genes interact in complex ways with a plurality of environmental factors, which might produce a genetic tendency toward criminal behaviour (or aggression, anti-social behaviour or violence). And yet – as captured aptly by Lyons' quote above – the choice between hard and soft determinism seems to be the choice between a theory that is clearly wrong and a theory that is probably right, but empty of content. To suggest that 'criminal' behaviour reflects an interplay of constellations of genes and complex environmental conditions does not seem far removed from suggesting that what people do (or what they become) reflects a relationship between persons and their context. The CIBA Symposium thus highlights what are perhaps the epistemic irreconcilables not only of biological and cultural standpoints, but of biology and culture in and of themselves. Yet, even as 'soft determinism' does not resolve the symposium's rather crude attempt to carve out a predictive calculus of crime and genes, it signals an important emergent counter-current within genetic science – a move toward complexity. As I shall discuss further below, the turns to genomics, epigenetics, and microbiomics (all presaged by, though not equivalent to, the 'soft determinist' framework) hold out significant potential to rupture the normative expectation, pre-eminent justification and normotic drive underpinning the plausibility of genes – that it will produce 'goods'.

Jewish Genes

A comparable example of biological conclusions based on cultural premises is to be found in the search for a 'Jewish gene'. This plays out through what might be termed a *circuit of presumptions*. As discussed in Chapter 7, the genetic marker sought in this context – from an African-Jewish community that was presumed not to be Jewish on their own account (and by reason of their being African and non-white) – was for the 'Kohanim' gene. Kohen being a patrilinial priestly status of Jews handed down from father to son, Kohanim have a specific role in rites and practices associated with worship. Aside from (and accruing to) its racial presumptions concerning what might be taken as a standard Jew (white and European), the quest for a Kohanim gene also rests on the prior assumption that cultural identity carries, and is reducible to, a biological marker which, and this is a second presumption, can be accurately pinpointed and traced by genetic means. In turn, there is the corollary presumption (in this case both of the science and of the documentary about it) that genetics can provide culturally–genealogically meaningful information – in this case, tell us something about Jewish identity, what it means to be a Jew, and the intersubjective relations of Jewish kinship. Given that Jews as a 'people' include a history of conversions and still-living communities that cross racial categories, the twin inference and assumption of biological coherence presents as a self-affirming tautology, disturbingly at odds with the actual, empirical history of Jewish people. Underpinning the science are two further presumptions: one being that 'Jewish' constitutes a coherent and plausible biological category, and the other, connected to this presumption, is that the category 'Kohen' (and the Lemba) can be assumed to be communities with unbroken lines of reproductive heritage. The notion that exact paternity back to antiquity can be *known* or *assumed to have followed* for the purpose of genetic genealogical testing is an extraordinary one, given what we know about the contemporary realities of human sexual relationships and the reliability of assumptions of paternity. If the logical inconsistencies of the search for Jewish genes in this instance suggest foundational epistemic problems with genetic genealogy as a field, they also suggest something of the character of its powerful cultural purchase. As with the example of genes and crime, identity genetics suggest a relationship between plausibility and promise. These encompass not only material 'goods' (nutritious and plentiful food, new cures for recalcitrant disease, definitive evidence of guilt or innocence) but more ephemeral and inchoate potentialities (a safer society, self-understanding, a place in history, belonging).

Tay-Sachs

If the genetics of crime and Jews seem informed by a tautological fallacy concerning the relationship of biology and culture (and the extent to which biological conclusions can be drawn from cultural categories), one could argue that the Tay-Sachs example does *not* do this. The indexical category – the disease process of Tay-Sachs – reflects a stable characterisation and consensus based on a sustained longitudinal and triangulated history of clinical experience and empirical observation by researchers, clinicians and parents. As signalled in the tone of the quote above, moreover, the Tay-Sachs example (and the website that elaborates on it) presents carefully delimited claims on that basis about the genetic biology of this condition. Yet at the same time, Tay-Sachs is also constituted on the terrain of culture. There are, for example, imperatives of action (or to use the terminology I suggested in Chapter 1, there is an *ethical burden*) that accrue(s) to the diagnostic framing of genetic disease. These are the normative expectations of *responsible* action that not only arise in the face of a positive pre-pregnancy or prenatal diagnosis (being a carrier or carrying an afflicted pregnancy respectively), but also frame the terms of prudence concerning which (prospective) parents should (be expected to) avail themselves of genetic diagnostic testing in the first place. The cited association of Tay-Sachs with Ashkenazi Jews is framed in careful and limited terms (a statistical association, not a characteristic and universal trait). It is clear, even in the short passage, that being Jewish does not mean one is a carrier of Tay-Sachs. Nor is it suggested that Tay-Sachs is only found among Jews. Yet the association nevertheless has a reciprocal effect, lending an aura of biological coherence to 'Ashkenazi' and 'Jew' in and of themselves, and this is troubling for many of the reasons emergent in the search for a Jewish gene. First, if one considers beyond the Tay-Sachs context, the history of Jewish identity does not seem to support an inference of biological continuity or coherent classification. The history of Jews is one that includes conversion (one does not need to be born a Jew or of Jews to be Jewish, but once converted or adopted, one *is* unequivocally a Jew), radical displacements and vast losses of whole communities with attendant destruction and loss of birth records (even if those records were accurate in the first place).[3] Second, the certainties of Tay-Sachs, both as a coherent disease category and one that is associated, however provisionally, not only with ethnic identity, but specifically with a Jewish identity, cannot help but also carry cultural resonances that have historically associated Jews with disease. Tay-Sachs is perhaps one of the strongest examples of genetic certainty. And yet, it has its uncertainties. It too, occupies a troubled terrain between biology and culture.

---

3 Historical–genealogical records of modern European Jewry have been lost through repeated expulsions (and the wider diaspora of the Jewish people from antiquity). Births also were not always recorded, even in relatively stable periods.

The BRCA Dilemma (and the Jolie Effect)

On 14 May 2013, the *New York Times* published an article by Angelina Jolie in which she disclosed that she had undergone a preventive double mastectomy after having been diagnosed with a BRCA 1 gene mutation and advised that she had an 87 per cent risk of developing breast cancer.[4] Jolie's decision and the ensuing torrent of media response were, in part, a testament to her capital, both as a celebrity and as 'the most beautiful woman in the world' – as publications across the print and digital media characteristically referred to her. But they also signalled the reciprocal freighting of genetic diagnosis in the context of cancer and the ways in which the impetus to action articulates with a wider array of medico-moral values, perhaps especially those surrounding female bodies. Her decision to take the most extreme surgical option – particularly as she was not diagnosed with cancer and also given that the calculus of risk that she referenced is against populations, *not* individuals[5] – was widely perceived, albeit with occasional disquiet, to be an edifying example of, as the quote above put it, 'genetic self-ownership'. That there were and are considerable uncertainties attending Jolie's decision, not least what it implies to validate – as an example of prudence and ethical self-responsibility – such radical, irreversible and damaging treatment of a healthy body because one might develop a disease, suggests a talismanic freighting of genes.

If cancer is one of the more persuasive contexts for genetic governmentality – with the latter's twin impetus to certainty and definitive action – cancer is also, perhaps, the most publicly prominent context in which competing paradigms of genes have emerged. On the one hand is the framework of the *single gene* (or gene mutation) taken as a marker of a propensity to disease, emblematised in the association between BRCA mutations and elevated risks of breast, ovarian and possibly uterine cancer. This is a paradigm that is particularly given to normotic expectation: that cause and effect can be pinpointed, that definitive action is possible (or required) and that scientific understanding will translate into clinical protocols. On the other hand are the turns to complexity represented by genomics (the study of whole genomes),

---

4 See Jolie, 14 May 2013. Jolie's mother had died of ovarian cancer and her maternal aunt, of breast cancer.

5 Risk as measured against populations does not provide a reliable indicator of risk for individuals. Women with BRCA 1 or 2 mutations do not necessarily go on to develop breast cancer. Most cancers are not associated with BRCA 1 or 2 mutations. A double mastectomy also does not obviate the possibility of developing breast cancer. See Konnikova (15 May 2013).

epigenetics (the study of the complex factors of gene expression)[6] and microbiomics (the study of the collective interaction – including genomic – of microbes).[7] These latter paradigms represent a significant departure from the monadic construction of genes and their potential effects (whether from a hard or soft determinist perspective).[8] They are, in effect, theories of interactionality – indeed of a gestalt of multiple interactional contexts from within and outside bodies – and as such, they provide at the very least, a troubled foundation for definitive action against risks (or actuality) of disease. Indeed, the epigenetics of cancer has progressively unravelled its central referent; 'cancer' is not one disease, but a multiplicity of conditions. Perhaps ironically, the logic of complex interactional fields of influence has produced a new discourse of action: genetically targeted (personalised) medicine. But this too offers a foundational uncertainty that threatens to unravel the guiding principles of medical treatment; how can one develop or sustain standards of care, or a sustainable political economy of drug or surgical intervention, without continuity of definition or a common referent? Targeting, as a concept, is paradigmatically singular. It posits controlled entry and effect. It presupposes the severability of the actions of treatment from the constitution of the field. Against the brutal blunt instrument of current cancer treatment, the projection of personalised treatment, geared to the genomic specificity of both cancer and patient and thus sparing patients unnecessary hardship and greater promise of cure, has obvious purchase. Yet *is* such a projection supported by these emergent paradigms – by the suggestion of genomic multiplicity, for example? What if a person's genomic profile might vary depending on the cells of origin that are tested? What if a cancer tumour in and of itself is genomically multiple (and mutating)?[9] As Jolie's decision and its wider public reception suggest, the persuasion of the singular gene paradigm is powerfully sedimented into public understanding of both cancer and genetics. And, indeed, the single gene framework continues to

6 These turns to complexity have both accrued to and produced re-evaluation of non-coding ('junk') DNA, formerly regarded as non-integral or without function but now understood as integral to (though not the only factors in play in) the 'switching' on and off of particular gene expressions (for a succinct summary of this 'turn' within genetics, see Kean, 3 October 2013).

7 Ninety per cent of the human body, for example, is composed of microbes and thus constitute a complex epigenetic and genomic ecology.

8 See, for example, the NOVA/PBS 2007 documentary *Ghost in Your Genes*, which explores the epigenetics turn in the context of ageing and cancer.

9 These are not abstract questions, of course, as genomic multiplicity has arisen in both the forensic (DNA analysis) context as well as in cancer research.

have purchase and prominence in scientific circles.[10] Investments (rhetorical and researched) in targeted medicine appear to bespeak of an emergent *war of position*;[11] not only over genetic scientific values (and the value of genetic science), but over the terrain of normotic expectation-attachment itself.

---

10 As I was completing this chapter, I happened to receive an email invitation through the University of Birmingham (UK) Alumni Office for a monetary donation to the University's programme of breast cancer research, specifically into BRCA 1 research:

Dear Deborah L. Steinberg

The research by me and my team at the University of Birmingham could mean future generations of women genetically predisposed to develop breast cancer will not have to live in fear or go through life-changing surgery. You can help us.

Emma's story

When Emma Parlons discovered she had a BRCA1 genetic mutation that meant she had an 85% chance of developing breast cancer, she could not imagine going through the emotional turmoil of being screened twice a year. Emma chose to have a double mastectomy.

'Screening is a hideous thing to prepare yourself for. For me, losing my breasts was a small sacrifice compared to having my life taken over by the prospect of getting cancer. Afterwards I was euphoric that I'd saved my life and avoided a hideous disease,' she says.

There is a 50% chance that Emma's nine-year-old daughter has the same genetic mutation, and Emma hopes research at the University of Birmingham will give more options to the next generation of BRCA1 carriers.

Our research

In my laboratory we are working towards making normal breast epithelial cells that have the same genetics as women like Emma. By making very precise changes in these cells so that they have particular BRCA1 mutations, we hope to discover the very early cellular changes that signal the cells are turning into cancer. Ultimately, we hope women might be treated for the very early signs of cancer without ever becoming too ill or having dramatic preventative treatments.

Dr. Jo Morris
Senior Lecturer, School of Cancer Sciences

[received October 11 2013]

These passages powerfully reinforce single gene (mutation) causality, while only alluding to a potential for interventions (left entirely unspecified) that will preclude current brutal options.

11 I refer here, albeit applied to a very different context, to Gramsci's concept of war of position, which refers to a hegemonic struggle of values (as distinct from contested tactics within a shared set of underlying values) (Gramsci 1971).

A Note on Dinosaur Parks

Jurassic Park, the central trope of Michael Crichton's (1990) book *Jurassic Park*, is a fiction. There has never been anywhere in the world a genetically resurrected dinosaur theme park. And yet, the 1993 film release *Jurassic Park* caught not only the larger popular imaginary, forging and dramatically amplifying the sedimentation of genes into cultural vernacular, but, in so doing, also leveraged the transformative purchase of the science of genetics. It is worth noting, for example, the existence of current and widely reported scientific research into the resurrection of the extinct woolly mammoth including a BBC Radio 4 documentary entitled *Raising Allosaurus: The Dream of Jurassic Park*.[12] But perhaps more to the point are the ways in which genes have come to constitute both referents and markers for a range of cultural values, perhaps most particularly, for the attribution of *authenticity*.

Gene and Episteme: Between Biology and Culture

The juxtaposition of the above examples powerfully evidence a reciprocal imbrication of biology and culture. The *circuits of meaning* surrounding genes reflect an inter-sedimentation of material and symbolic relations, of languages and bodies, of values and governance, of knowledge and feeling. They also suggest a schism between *factuality* and *reality*, that the facts of genes are less stable, less logically coherent, less empirically grounded, less *important,* than one might have expected. And yet the *reality* of genes is pervasive. Indeed, one might posit that the reality of genes lies less in their factuality than in the modes of attachment that surround them.

One mode of attachment of course arises from the material relations, political economies and investment trends of science. Genes and genetics are pre-eminent sites of economic investment and monetisation across both private and public economies. Genes and genetics are, moreover, at the interstices of technological transformations from medicine and health, to agriculture, to pharmaceuticals, to bio-weapons and to informatics. Genes are loci of capital; they constitute currencies of exchange; they are sites of governance and regulation; they are an *order of things*.[13]

A second locus of attachment is commonsense and the ways in which genetic discourse reinvests in the terrain of the 'already known' and 'fits' the

---

12 *Raising Allosaurus: The Dream of Jurassic Park* was broadcast on BBC Radio 4 (30 August 2013). See also, for example, Sample (31 July 2013).

13 Here I am in part referring to Foucault's 1966 text by that title in which he outlines the concept of the 'episteme' which has been explored (and expanded) in this book.

already existing social order and prevailing values of the culture. Hence, the justification for querying the Lemba people is based on already understood (uncomfortable) associations between Jews and race and normative expectations about which bodies need explanation and which do not, and who has the authority to make those explanations. A third order of attachment arises, as we see in the Tay-Sachs and BRCA examples, from the *ethical burden* of genes – from the imperatives of action, the governmentality (including its unspoken, taken for granted implications in terms of gender) – that accrues to genetic knowledge.

Still another order of attachment is affective. Genes are – demonstrably, pervasively – points of projection for a potent array of phantasies, anxieties and desires present in and across cultures.

Phantasmatics and the Gene

Perhaps the most potent of the phantasmics of genes is normotic. This is a deep investment in the notion that we can know what is unknown and by that knowledge, control that which appears to be uncontrollable. This desire underpins investments in the *predictive rationality* of genetics, in the notion that genetic science can – it is only a matter of time – not only accurately map human bodies (all bodies), but can tell us what we need to understand about human problems, issues or aspirations so that we might act to control, prevent or cultivate them. It is in the frame of normotic desire that genes have emerged as plausible explanatory theories applied to disparate phenomena that range from identification of diseases of the body, to social ills, to cultural identity. Normotic phantasy underpins what might be termed the *fallacy of more* – that is, the notion that designated solutions will inevitably and logically arise from the accumulation of knowledge. It is a fallacy that bridges (and confuses) imagined potentialities (targeted medicine, predictions of criminality, enhanced abilities, the resurrection of woolly mammoths) and actual effects. Normotic science refuses the possibility that more knowledge might negate its starting paradigm, might lead away from definitive modes of action, might confuse or complicate decisions, and indeed is contradicted by the Popperian understanding of science as characterised by disprovability. The science that arises out of normotic phantasy cannot envision (and would repudiate) being disproved and thus learned from – it is driven only to 'being used'. One consequence of normotic projection is constraint on the conditions of research, positioning emergent knowledges that mitigate *against* action as failure, as undesirable, as what we cannot bear and do not want to know. This is the particular challenge of (and for) the 'turns' to complexity in genetics – they suggest uncertainties, ambivalence – both in the character of bodies and the possibilities of diagnosis, care or cure.

They do not seem to lend themselves easily to the corollary normotic phantasy – the *phantasy of action,* with its defence against loss, and its index of what is viable, ethical, or necessary. The discourse of 'personalised' genetic medicine – emergent out of and yet at odds with the anti-normotic potentialities of epigenetic and genomic 'turns' – seems founded on the principle that bodily complexity can be contained, can be resolved, can be selectively effected. Such a proposition is in many respects over-determined by the normative political economies of science that demand, as an index of research value, utility or translation into deliverable, marketisable products, services, answers, action. What happens when more knowledge unravels its referent – when it undoes cancer or the DNA fingerprint as meaningful categories for example, or when it suggests that we are composed of multiple genomic signatures? The notion that progress is linear, evolutionary, and always resolves to action denies the history of science.

A second order of attachment is the *phantasy of belonging.* This is the primary desire – existential and intersubjective – for communion, for recognition attendant on the revelation of one's authentic self. The gene in the context constitutes a metonym of historical location, to the entitlements that follow belonging, including the *filling in* of the empty spaces of self, of the lost connection. It is also founded on a powerful mode of transference in which a powerful knower can know you, can define your possibilities, can locate you in a place to which you unequivocally belong, can place you unequivocally in a lineage of genealogical inheritance, and at its most grand – in the sweep of human history. The gene crystallises this transference, is at once its origin myth and its artefact.

Then there is the *phantasy of transcendence.* At the immediate level, this is the notion that there is within us, a redemptive – and because it is of the body, unarguable – inner personhood that can be found and freed, an authentic 'I' that will bring with it a life replete with meaning, connected to histories beyond one's own self. It is a corollary investment of faith that genetics (that science) can locate, can identify, can set in motion, can authorise (can even author) an inner embodied possibility to transcend limitation, whether this is overcoming disease, cultivating desired capabilities, finding our 'true' histories, or correcting social wrongs. At a meta-level, this phantasy underscores genetics as a redemptive, remedial project against the historical wrongs of science itself.

Orders of Real

Genes are subject to many orders of reality. They are points of convergence for a vast constellation of scientific, medical, educational, political, penal, economic and cultural institutions. Genes are sites of specialist language, of expert knowledges and practices applied to living matter. They are an explanatory vocabulary – of kinship, social identity of history, of futurity. Genes are objects of representation: in peer-review, the news, film, and fiction. They are objects of desire, or repudiation.

Ultimately, genes are also sites of excess. If factuality is figured as a correspondence between a concept and its material referent, the evidence suggests that what is real about genes is in excess of whatever may be said to be factual about them.[14] One might almost say that the fictions of genes appear to be their primary reality. If we consider the claims of fact about genes, we find premises and inferences that do not seem to hold up under scrutiny. Very often these are cultural assumptions or categorisations that are themselves shaky, that are moving targets, emerging at particular moments in time, and points of debate and contestatory construction. We find an inter-permeability between science and culture that makes it difficult to disentangle facts from fiction, and that suggest that facts may be more an effect of desire than they are of a coherent, empirically grounded programme of research or observation. And we find consequences that are equally problematic. On scrutiny, genes appear persuasive not because of their proven facts, but because of the profound gaps in knowledge – their empty spaces – into which so much, it seems, can be projected. From this vantage point, the empirically established facts about genes are only a piece of the reality of genes, indeed, arguably, the least consequential piece.

We might conclude that a biological standpoint tells us less than we might think and a cultural standpoint tells us more than we might think about the reality of genes. As the case studies in this book suggest, the biological facts of genes are often attached to the most cloudy of concepts as if there were a one-to-one correspondence. Even with the most seemingly stable correspondence between genetic concepts and their empirical referents, the facts begin to crumble. I am not suggesting here that biology and culture are opposing camps, or that the cultural realities of the gene constitute an (invalid) overlay to its biology. Rather, I am suggesting that culture – including science – forms the context, locus and foundation of the search for genes. Genes are the reciprocal work of science and culture, and of science as culture. I am also not suggesting that the instability of genetic facts means that genes are not real. It strikes me that genes are all too real.

---

14 As aptly suggested by Dobbs in his *Slate* article 'Genetics' Rite of Passage' (27 October 2013) which questioned both the factual stability and (dearth of) empirical pay off of genetics research: 'Find a gene, slap high-fives around the lab, publish to huzzahs, and then ... take it all back'.

(Un)Beautiful Gene

I have an antipathy to genes, a profound phantasmatic desire to repudiate them.

This is a somewhat ironic point, since I have just argued at length that there is a cultural investment in genes that seems deeply ingrained notwithstanding whether the facts actually justify that faith. And now I am saying that I want to refuse genes no matter what the facts are. My disaffection for genes arises in part from the instabilities or illogics I perceive in the characterisation of genes as facts. But it is more than this; it is because the conditions of genes – the realities that produced the knowability of genes and that in turn, genes produce – disturb me. Perhaps because of my peculiar mode of cultural standpoint and my understanding of what is real, I both believe in and am invested in qualified facts. I feel resistant to a world order that is genetic; it seems to me that there is a foundational dissonance between the redemptive narratives attached to genes and their governmentality. I do not want a government of genes. And I cannot seem to imagine an alternative that does not recuperate back to normotic phantasy and normative inequalities. I find that I do not want genes to be real.

In *The End of Faith*, Sam Harris (2006) posits the notion of an 'ethical fact', by which he means, in part, the call to interrogate the ethical consequences of facts. But more, this suggests the intextricability of facts from judgements of value. This is an interesting and also disturbing proposition. On the one hand it appears to align bias with (and as an intrinsic quality of) both knowledge and science. It refutes the possibility of a search for knowledge that is not inextricably intertwined with values, that is not in some sense *normative* (even if subversively or alternatively so). The conventional notion of an 'ethical fact' implies its obverse – the 'unethical fact'. Both notions understand knowledges as *tendencies* – emergent in their foundational assumptions and accruing to the conditions of their pursuit, the contexts of their dissemination, the languages through which they circulate. In this sense, knowledge is always ethically charged, as it constitutes not only an interpretive, but an *imperative* field. At its (arguably) crudest, this can be seen in both the search for, and findings of, genetic factors in criminality, BRCA mutations, or Jewish genes – all of which *action* the world in some sense, demand a re-ordering of things, a balancing, a summons of state or individual to judge. An 'ethical fact', according to Harris, might be one charged with the imperatives of justice, or social goods, or the Levinasian ethic of feeling for the other. Yet one of the features of the genetic landscape is that genes encompass contrary notions of ethics: eugenic and anti-racist; a government of genes; a promise of self; a reduction of history; a beautiful journey.[15]

---

15 A case in point is the framing of Lone Frank's (2010) book: *My Beautiful Genome: Exposing our Genetic Future, One Quirk at a Time.*

Genes and the science of genetics are clearly embedded in a social world: arising from distinctive, historically specific social conditions, and giving rise to social consequences. The apotheosis of the gene challenges traditional views of scientific enterprises as detached from or superordinate to its social context, and of truth as distinct from value. Biology, culture; genes are artefacts of both. And they are indeed, as Steve Jones put it – although not, perhaps as he meant it – a language. They represent not only the 'grammar' of biochemical interaction, but equally significantly, the conversation between cultural significance and the preconditions of scientific imagination.

The cultures of the gene evidence science as, among the many things it is, a culture industry. It is as much an endeavour of communications and signification as it is of empirical labours in the lab or in the field. Genes demonstrate the place of science not only in technological application, but in the building of worlds, in the cultural imaginary, in questions of identity, love, alienation, and aspiration. In this light, to question the possibility of an ethical gene, even to wish the gene undone, is more than a rhetorical exercise. It is to join its science. It is, ironically perhaps, to articulate an alternative way not only to read, but also to author the world, through genes.

Bibliography

Anderson, B. 1982. *Imagined Communities*. London: Verso.

Anker, S. and D. Nelkin. 2003. *The Molecular Gaze: Art in the Genetic Age*. New York: Cold Springs Harbor Laboratory Press.

Anthias, F. and N. Yuval-Davis. 1993. *Racialized Boundaries: Race, Nation, Gender, Colour and Class and the Anti-Racist Struggle*. London: Routledge.

Arendt, H. 1963 [2006]. *Eichmann in Jerusalem*. London: Penguin Classics.

Barker, M. 1981. *The New Racism: Conservatives and the Ideology of the Tribe*. London: Junction Books.

Barr, M.S. 2000. *Future Females: The Next Generation: New Voices and Velocities in Feminist Science Fiction Criticism*. Lanham, MD: Rowman and Littlefield Publishers.

Barr, M.S. 2006. *Lost in Space: Probing Feminist Science Fiction and Beyond*. Chapel Hill: University of North Carolina Press.

Barthes, R. [1957] 1972 edition. *Mythologies* (trans. Annette Lavers). New York: Hill and Wang.

Beck, U. 1992. *Risk Society: Towards a New Modernity*. London: Sage.

Benjamin, J. 1988. *The Bonds of Love: Psychoanalysis, Feminism and the Problems of Domination*. New York. Pantheon.

Benston, M.L. 1992. 'Women's Voices/Men's Voices: Technology as a Language' in G. Kirkup and L. Smith Keller, *Inventing Women: Science, Technology and Gender*. London: Polity Press, pp. 33–41.

Birke, L. 1994. 'Zipping up the Genes; Putting Biological Theories Back in the Closet', *Perversions* 1: 38–51.

Boler, M. 2006. 'Uses of Multimedia for Political Dissent', paper presented at Biopolitics and Technoscience workshop, 30 November, University of Toronto.

Boling, P. (ed.) 1995. *Expecting Trouble: Surrogacy, Fetal Abuse and New Reproductive Technologies*. Boulder, CO: Westview Press.

Bollas, C. 1987. *The Shadow of the Object: Psychoanalysis of the Unthought Known*, London: Free Association Books.

Bordo, S. 1993. *Unbearable Weight: Feminism, Western Culture and the Body*. Berkeley: University of California Press.

Brah, A. and A. Coombs. 2000. *Hybridity and its Discontents: Politics, Science, Culture*. London: Routledge.

Braudy L. and M. Cohen (eds). 2009. *Film Theory and Criticism,* seventh edition. New York: Oxford University Press.

Bristow, J. 1991. *Empire Boys: Adventures in a Man's World.* London: HarperCollins.

British Broadcasting Corporation (BBC). 1993. 'Producers' Guidelines' in J. Corner and S. Harvey (eds), *Television Times: A Reader.* London: Arnold, pp. 246–52.

Brookley, R.A. 2002. *The Rhetoric and Power of the Gay Gene.* Bloomington: Indiana University Press.

Burdekin, K. 1985 [1937]. *Swastika Night.* London: Lawrence and Wishart.

Burton, A. 1990. 'The White Woman's Burden: British Feminists and the Indian Woman 1865–1915', *Women's Studies International Forum* 13(4).

Butler, J. 1993. 'Endangered/Endangering: Schematic Racism and White Paranoia' in R. Gooding-Williams (ed.), *Reading Rodney King: Reading Urban Uprising.* New York: Routledge, pp. 15–23.

Butler, J. 1993. *Bodies that Matter: On the Discursive Limits of Sex.* New York: Routledge.

Carradine, E. 2008. *Crime, Culture and the Media.* Cambridge: Polity Press.

Chambers English Dictionary. 1990. Edinburgh: Chambers.

CIBA Foundation Symposium 194. 1996. *Genetics of Criminal and Antisocial Behaviour (Symposium 194).* Chichester: John Wiley & Sons.

CIBA Foundation Symposium. 1986. *Human Embryo Research: Yes or No?* London: Tavistock.

Clarke, A. (ed.). 1994. *Genetic Counselling: Practice and Principles.* London: Routledge.

Clarke, C.A. 1987. *Human Genetics and Medicine (Third Edition).* London: Edward Arnold.

Cohan, S. 2008. *CSI: Crime Scene Investigation.* London: Palgrave.

Cohen, S. 1972. *Folk Devils and Moral Panics: The Creation of Mods and Rockers.* London: MacGibbon & Kee.

Colker, R. 1994. *Pregnant Men: Practice, Theory and the Law.* Bloomington: Indiana University Press.

Collins, P. Hill. 1990. *Black Feminist Thought: Knowledge, Consciousness, and the Politics of Empowerment.* Boston, MA: Unwyn Hyman.

Colls, R. and P. Dodd. 1986. *Englishness: Politics and Culture 1880–1920.* London: Croom Helm.

Connor, S. and T. Wilkie. 1993. 'The Gay Gene', *Independent*, 18 July, p. 8.

Connor, S. 1993. 'Gay Gene Raises Host of Issues', *Independent*, 16 July, p. 3.

Connor, S. 1993. 'Homosexuality Linked to Genes', *Independent*, 16 July, p. 1.

Connor, S. 1993. 'Elusive Answers to Darwin's Riddle', *Independent*, 17 July, p. 3.

Connor, S. and M. Whitfield. 1993. 'Scientists at Odds in "Gay Gene" Debate', *Independent*, 17 July, p. 3.

Cook-Deegan, R. 1994. *The Gene Wars: Science, Politics, and the Human Genome.* New York: W.W. Norton & Company.

Cooper, M. 2008. *Life as Surplus: Biotechnology and Capitalism in the Neoliberal Era*. University of Washington Press.

Corea, G. 1979. *The Hidden Malpractice: How American Medicine Mistreats Women*. New York: Harper and Row.

Corner, J. and S. Harvey (eds). 1993. *Television Times: A Reader*. London: Arnold.

Crichton, M. 1991. *Jurassic Park*. London: Random House.

Crichton, M. 1995. *The Lost World*. London: Random House.

Crosby, J.C. 2000. *Cauldron of Changes: Feminist Spirituality in Fantastic Fiction* Jefferson, NC: McFarland and Co. Inc.

Curran, J. and J. Seaton. 1988. *Power Without Responsibility: The Press and Broadcasting in Britain (Third Edition)*. London: Routledge.

Daily Mail. 1993 'Scan Fears', 16 July, p. 8.

Davis, A. 1990. 'Racism, Birth Control and Reproductive Rights' in M.G. Fried (ed.), *From Abortion to Reproductive Freedom: Transforming a Movement*. Boston, MA: South End Press, pp. 15–26.

Dawson, G. 1994. *Soldier Heroes: British Adventure, Empire and the Imagining of Masculinities*. London: Routledge.

Denno, D.W. 1996. 'Legal Implications of Genetics and Crime Research' in CIBA Foundation Symposium 194, *Genetics of Criminal and Antisocial Behaviour (Symposium 194)*. Chichester: John Wiley & Sons.

Dobbs, D. 2013. 'Genetics' Rite of Passage'. *Slate*, 27 October <http://www.slate.com/articles/health_and_science/human_genome/2013/10/human_genetics_successes_and_failures_ashg_stories_of_disease_genes.html>.

Douglas, M. 1966. *Purity and Danger: An Analysis of the Concepts of Pollution and Taboo*. London: Routledge and Kegan Paul.

Doyle, B. 1986. 'The Invention of English' in R. Colls and P. Dodd, *Englishness: Politics and Culture 1880–1920*. London: Croom Helm.

Du Gay, P., S. Hall, L. Janes, A. Koed Madsen, H. Mackay and K. Negus. 1997. *Doing Cultural Studies: The Story of Sony Walkman*. Milton Keynes/Thousand Oaks, CA: Open University/Sage.

Dubow, S. 2010. *Ourselves Unborn: A History of the Fetus in Modern America*. Oxford: Oxford University Press.

Elkington, J. 1985. *The Poisoned Womb: Human Reproduction in a Polluted World*. Harmondsworth: Penguin.

Epstein, D. and D.L. Steinberg. 2007. 'The Face of Ruin: Evidentiary Spectacle and the Trial of Michael Jackson', *Social Semiotics* 17(4): 441–58.

Epstein, D. and J.T. Sears. 1999. *A Dangerous Knowing: Sexuality and the 'Master Narrative'*. London: Cassell.

Epstein, D., R. Johnson and D.L. Steinberg. 2000. 'Twice Told Tales: Re-Telling Sexual Stories in the Age of Consent Debates 1998', *Sexualities* 3(1): 5–30.

Epstein, D. 1993. *Changing Classroom Cultures: Anti-Racism, Politics and Schools*. Stoke-on-Trent: Trentham Books.

Epstein, D. 1997. 'What's in a Ban: The Popular Media, Romeo and Juliet and Compulsory Heterosexuality' in D.L. Steinberg, D. Epstein and R. Johnson (eds), *Border Patrols: Policing the Boundaries of Heterosexuality*. London: Cassell, pp. 183–203.

Epstein, J. 1995. 'The Pregnant Imagination, Fetal Rights and Women's Bodies: A Historical Inquiry', *Yale Journal of Law and the Humanities* 7(1).

Ettorre, E. 2002. *Reproductive Genetics, Gender and the Body*. London: Routledge.

Ettorre, E. 2006. 'Re-Shaping the Space Between Bodies and Culture: Embodying the Biomedicalised Body', *Sociology of Health and Illness* 20(4): 548–55.

Ettorre, E. 2007, *Genetics, Gender and an Embodied Ethics of Reproduction*, presented at CESAGen Seminar, March, Lancaster University.

Ettorre, E., B. Katz Rothman and D.L. Steinberg. 2006a. 'Feminism Confronts the Genome: Introduction', in Ettorre, E. et al. (guest eds), 'Feminism Confronts the Genome' (Special Issue) *New Genetics and Society* 25(2): 133–42.

Ettore, E., B. Katz Rothman and D.L. Steinberg (guest eds). 2006b. 'Feminism Confronts the Genome' (Special Issue) *New Genetics and Society* 25(2).

Ewing, C. 1995. 'A Gay Gene? A Critique of Genetic Research of Homosexuality', Unpublished paper, presented at the Intercampus Gay and Lesbian Seminar Group, La Trobe University, Melbourne, Australia, 5 September.

Fausto-Sterling, A. 1985. *Myths of Gender: Biological Theories about Women and Men*. New York: Basic Books.

Feinman, C. (ed.) 1992. *The Criminalization of a Woman's Body*. New York: Harrington Park Press.

Fenton, J. 1993. 'It's not in the Genes, it's in the Culture', *Independent*, 19 July, p. 19.

Foucault, M. 1966. [2002 edition] *The Order of Things: An Archaeology of the Human Sciences*. London: Routledge.

Foucault, M. 1977. *Discipline and Punish: The Birth of the Prison*. London: Penguin.

Fox Keller, E. 1984. *A Feeling for the Organism: The Life and Work of Barbara McClintock*. New York: W.H. Freeman.

Keller, E. Fox. 2000. *The Century of the Gene*. Maryland University Press.

Frank, L. 2010. *My Beautiful Genome: Exploring Our Genetic Futures: One Quirk at a Time*. Oxford: Oneworld Publications.

Franklin, S., Geesink, I. and Prainsack, B. 2008 (guest eds). *Stem Cell Stories 1998–2008*. Special Issue of *Science as Culture* 17(1).

Franklin, S. and C. Roberts. 2006. *Born and Made: An Ethnography of Preimplantation Genetic Diagnosis*. New York: Princeton University Press.

Franklin, S., Edwards, J. et al. 1999. *Technologies of Procreation: Kinship in the Age of Assisted Conception, 2nd edition*. London: Routledge.

Franklin, S. 1991. 'Fetal Fascinations: New Dimensions to the Medical-Scientific Construction of Fetal Personhood' in S. Franklin, C. Lury, and J. Stacey (eds), *Off-Centre: Feminism and Cultural Studies*. London: Harper Collins.

Franklin, S. 1997. *Embodied Progress: A Cultural Account of Assisted Conception.* London: Routledge.

Freidson, E. 1970. *Profession of Medicine: A Study of the Sociology of Applied Knowledge.* New York: Dodd, Mead.

Freire, P. 1970. *Pedagogy of the Oppressed.* New York: Seabury.

Fyfe, W. 1991. 'Abortion Acts: 1803 to 1967' in S. Franklin, C. Lury, and J. Stacey (eds), *Off-Centre: Feminism and Cultural Studies.* London: HarperCollins.

Garrett, L. 1994. *The Coming Plague: Newly Emerging Diseases in a World Out of Balance.* New York: Penguin.

Gilman, S., 1985. 'Black Bodies, White Bodies: Toward an Iconography of Female Sexuality in Late Nineteenth-Century Art, Medicine, and Literature', in H. Gates (ed.), *Race, Writing and Difference.* Chicago: University of Chicago Press, pp. 223–61.

Gilman, S. 1991. *The Jew's Body.* New York: Routledge.

Gilman, S. 1999. *Making the Body Beautiful: A Cultural History of Aesthetic Surgery.* New York: Princeton University Press.

Goodwin, A. and G. Whannel (eds). 1990. *Understanding Television.* London: Routledge.

Gottweiss, H. 1997. 'Genetic Engineering, Discourses of Deficiency, and the New Politics of Population' in P. Taylor, et al. (eds), *Changing Life, Genomes, Ecologies, Bodies, Commodities.* Minneapolis: University of Minnesota Press, pp. 56–84.

Gover, J. 1996. 'The Implications for Responsibility of Possible Genetic Factors in the Explanation of Violence' in CIBA Foundation Symposium 194, *Genetics of Criminal and Antisocial Behaviour (Symposium 194).* Chichester: John Wiley & Sons.

Graddol, D., D. Leith and J. Swann (eds). 1996. *English: History, Diversity and Change.* London: Routledge/Open University.

Gramsci, A. 1971. *Selections from the Prison Notebooks* (eds and trans. Q. Hoare and G. Nowell-Smith). New York: International Publishers.

Gray, A. 1992. *Video Playtime: The Gendering of Leisure Technology.* London: Routledge.

Gregory, R. 2012. 'A Womb With a View: Identifying the Culturally Iconic Fetal Image in Prenatal Ultrasound Provisions', *Present Tense: A Journal of Rhetoric in Society* 2(2).

Grosz, E. 1994. *Volatile Bodies: Towards a Corporeal Feminism.* Bloomington: Indiana University Press.

Guardian. 1993. 'Science and the Single Man', (Outlook Section), 17 July, p. 1.

Guardian. 1993. 'Opting for a Gay Child', (Letters to the Editor), 17 July, p. 17.

Habgood, J. 1993. 'We Are All People, Not Gene Models'. *The Times*, 24 July, p. 7.

Hall, S. et al. 1978. *Policing the Crisis; Mugging, the State, and Law and Order*. London: Macmillan.

Hall, S. 1982. 'Culture and the State' (Unit 28) of *The State and Popular Culture* (Block 7) of *Popular Culture* (Open University Course U203). Milton Keynes: Open University Press.

Halliday, S. and D. Lynn Steinberg (guest eds). 2004. *Medical Law Review Special Issue* 'The Regulated Gene: New Legal Dilemmas'. Spring, 12(1).

Hamer D.H., S. Hu, V.L. Magnuson, N. Hu and A.M. Pattatucci. 1993. 'A linkage between DNA markers on the X chromosome and male sexual orientation', *Science* 261(5119): 321–7.

Hamer, D.H. and P. Copeland. 1994. *The Science of Desire: The Search for the Gay Gene and the Biology of Behavior*. New York: Simon & Schuster.

Haran, J. 2003. *Re-Visioning Feminist Futures: Literature as Social Theory*, PhD Thesis, University of Warwick.

Haran, J. 2004. 'Theorizing (Hetero)sexuality and (Fe)male Dominance', *Extrapolation* 45(1): 89–102.

Haran, J. et al. 2007. *Human Cloning in the Media: From Science Fiction to Science Practice*. London: Routledge.

Haraway, D.J. 1989. *Primate Visions: Gender, Race and Nature in the World of Modern Science*. London: Verso.

Haraway, D.J. 1991. *Simians, Cyborgs and Women: The Reinvention of Nature*. London: Free Association Books.

Haraway, D.J. 1997. *Modest Witness @ Second Millennium: FemaleMan Meets OncoMouse*. New York: Routledge.

Harding, S. (ed.). 1993. *The 'Racial' Economy of Science: Toward a Democratic Future*. Bloomington: Indiana University Press.

Harris, S. 2006. *The End of Faith: Religion, Terror and the Future of Reason*. New York: Free Press.

Hawkes, N. 1993. 'Is Homosexuality all in the Brain?', *The Times*, 8 July, p. 15.

Hawkes, N. 1993. 'Tampering with the Bad Temper Gene', *The Times*, 1 July, p. 15.

Hawkes, N. 1993. 'Gays may have Genetic Link', *The Times*, 16 July, p. 3.

Hawkes, N. 1993. '"Gay Gene" Raises Screening Fear', *The Times*, 17 July, p. 3.

Hellekson, K. 2000. 'Toward a Taxonomy of the Alternate History Genre', *Extrapolation* 41(3): 248–58.

Herbert, F. 1976. *The Eyes of Heisenberg*. London: New English Library.

Hirt-Manheimer, A., J. Weinberg and J. Entine. 2008. 'Cracking the Code: DNA Detectives Reveal Secrets of the Jewish Past', *Reform Judaism*, pp. 28–36.

Hodson, A. 1992. *Essential Genetics*. London: Bloomsbury.

Holmes, R. 2006. *The Hottentot Venus*. Bloomsbury: Random House.

hooks, b. 1982. *Ain't I a Woman: Black Women and Feminism*. London: Pluto.

Hope, A. 1993. 'Scan Babies "More Often Left-Handed"', *Daily Mail*, 16 July, p. 26.

Hope, J. 1993. 'Genes that may Chart Course of a Sex Life', *Daily Mail*, 17 July, p. 20.

Hubbard, R. and E. Wald. 1993. *Exploding the Gene Myth: How Genetic Information is Produced and Manipulated by Scientists, Physicians, Employers, Insurance Companies, Educators and Law Enforcers*. Boston, MA: Beacon Press.

Hubbard, R. 1990. *The Politics of Women's Biology*. New Brunswick and London: Rutgers University Press.

Hundt, G.L., J. Sandall, K. Spencer, B. Heyman, C. Williams, R. Grellier, L. Pitson and M. Tsouroufli. 2008. 'Experiences of first trimester antenatal screening in a one-stop clinic', *British Journal of Midwifery* 16(3): 156–9.

Huxley, A. 1932. *Brave New World*. London: Chatto & Windus.

Illich, I. 1976 [1985 edition]. *Limits to Medicine; Medical Nemesis – The Expropriation of Health*. Harmondsworth: Penguin.

Illman, J. 1989. 'Cell Test has Key to Future', *Guardian*, 1 March.

Independent. 1993. 'Gay Genes do not Exclude Choice', 16 July, p. 19.

Independent. 1993. 'A Controversial Chromosome' (Letters to the Editor), 17 July, p. 15.

Independent. 1993. 'The Genetic Tyranny', 18 July, p. 24.

Independent. 1993. 'Nothing New in Predicting Sexual Orientation' (Letters to the Editor), 19 July, p. 17.

Irigaray, L. 1985. *Speculum of the Other Woman* (trans. G.C. Gill). Ithaca: Cornell University Press.

Jagger, A. and S. Bordo. *Gender/Body/Knowledge: Feminist Reconstructions of Being and Knowing*. New York: Rutgers University Press.

Johnson, R. and D.L. Steinberg (eds). 2004. *Blairism and the War of Persuasion: Labour's Passive Revolution*. London: Lawrence and Wishart.

Johnson, R. 1986. 'The Story So Far and Further Transformations?' in D. Punter (ed.), *Introduction to Contemporary Cultural Studies*. London: Longman.

Johnson, R. 1986–7. 'What is Cultural Studies Anyway', *Social Text* 16: 38–80.

Jones, D. 2006. 'The Science of You: The Ultimate Guide to Self Knowledge – Genes, Ancestry, Personality, Health, Intelligence, Beliefs' *New Scientist*, 19 August.

Jones, J.H. 1981 [1993 expanded edition]. *Bad Blood: The Tuskegee Syphilis Experiment*. New York: The Free Press.

Jones, S. 1991. *The Language of the Genes (The 1991 Reith Lectures)*. Transcripts. London: BBC Radio, Science Unit.

Jones, S. 1993. *The Language of the Genes: Biology, History and the Evolutionary Future*. London: Flamingo.

Jones, S and B. Van Loon. 2011. *Introducing Genetics*. London: Icon Books.

Jones, S. 1993. 'Genes, Gays and a Moral Minefield', *Daily Mail*, 17 July, p. 8.

Jones, S. 1997. *In the Blood: God, Genes and Destiny*. London: Flamingo.

Jones, S. 2002. *Y: The Descent of Men*. London: Little Brown.

Jordanova, L. (ed.). 1986. *Languages of Nature: Critical Essays on Science and Literature*. London: Free Association Books.

Jordanova, L. 1986. 'Introduction' in L. Jordanova. *Languages of Nature: Critical Essays on Science and Literature*. London: Free Association Press.

Kaplan, G. and L.J. Rogers. 2003. *Gene Worship: Moving Beyond the Nature/Nurture Debate over Genes, Brain and Gender*. New York: Other Press.

Kear, A. 1997. 'Eating the Other: Imaging the Fantasy of Incorporation' in D.L. Steinberg, D. Epstein and R. Johnson, *Border Patrols: Policing the Boundaries of Heterosexuality*. London: Cassell, pp. 253–74.

Keller, E. Fox. 2000. *The Century of the Gene*. Cambridge, MA: Harvard University Press.

Kevles, D.J. 1985. *In the Name of Eugenics: Genetics and the Uses of Human Heredity*. Harmondsworth. Penguin.

King, M.L. Jr. (James Melvin Washington ed.). 1986. *I Have a Dream: Writings and Speeches that Changed the World*. San Francisco: HarperCollins.

Kipling, R. 1999 [1888] 'The Man Who Would be King' in R. Kipling and L.L. Cornell, *The Man Who Would be King and Other Stories*. Oxford: Oxford University Press.

Kirkup, G. and L. Smith Keller. 1992. *Inventing Women: Science, Technology and Gender*. Cambridge: Polity Press.

Kirkup, G. et al. (eds), 2000. *The Gendered Cyborg: A Reader*. New York: Routledge.

Kitcher, P. 1996. *The Lives to Come: The Genetic Revolution and Human Possibilities*. London: Penguin.

Klein, R. and R. Rowland. 1988. 'Women as Test-Sites for Fertility Drugs: Clomiphene Citrate and Hormonal Cocktails', *Reproductive and Genetic Engineering: Journal of International Feminist Analysis* 1(3): 251–74.

Koonz, C. 1987. *Mothers in the Fatherland: Women, the Family and Nazi Politics*. New York: St. Martin's Press.

Kramer, A.-M. 2011a. 'Mediatizing memory: history, affect and identity in "Who Do You Think You Are?"', *European Journal of Cultural Studies* 14(4): 428–45.

Kramer, A.-M. 2011b. 'Kinship, affinity and connectedness: exploring the role of genealogy in personal lives', *Sociology* 45(3): 379–95.

Kranko, C. 2002. 'What-if-ing the Titans: Nancy Kress's Dialogic Beggars Trilogy', *Extrapolations* 43(2): 131–63.

Kress, N. 1993. *Beggars in Spain*. New York: HarperCollins.

Kress, N. 1994. *Beggars and Choosers*. New York: Tom Doherty Associates.

Kress, N. 1996. *Beggars Ride*. New York: Tom Doherty Associates.

Kristeva, J. 1992. *Black Sun: Depression and Melancholia* (trans. L.S. Roudiez). New York: Columbia University Press.

Larbalestier, J. 2006. *Daughters of Earth: Feminist Science Fiction in the Twentieth Century*. Middletown: Wesleyan University Press.

Latour, B. 1987. *Science in Action*. Cambridge, MA: Harvard University Press.

Lefanu, S. 1988. *In the Chinks of the World Machine: Feminism and Science Fiction* London: The Women's Press.

Lemke, T. 2003. 'Risk as Responsibility – Genetic Diagnosis, Moral Obligations and Consumer Choices' (Paper presented at the Vital Politics conference 5–7 September 2003, London School of Economics and Political Science. <http://www.thomaslemkeweb.de/engl.%20texte/Risk%20as%20Responsibility%20III.pdf>.

Lemke, T. 2004. 'Disposition and determinism – genetic diagnostics in risk society', *The Sociological Review* 52: 550–66 <http://www.thomaslemkeweb.de/engl.%20texte/Disposition.pdf>.

Lifton, R.J. 1986. *The Nazi Doctors: Medical Killing and the Psychology of Genocide.* New York: Basic Books.

Little, J.A. 2007. *Feminist Philosophy and Science: Utopias and Dystopias.* Amherst, New York: Prometheus Books.

Long, P. and T. Wall. 2009. *Media Studies: Texts, Production and Context.* Harlow: Pearson Education Ltd.

Lupton, D. 1994. *Medicine as Culture: Illness, Disease and the Body in Western Societies.* London: Sage.

Lyons, M.J. 1996. 'A Twin Study of Self-Reported Criminal Behaviour' in CIBA Foundation Symposium 194, *Genetics of Criminal and Antisocial Behaviour (Symposium 194).* Chichester: John Wiley & Sons.

MacCabe, C. and O. Stewart (eds). 1986. *The BBC and Public Service Broadcasting.* Manchester: Manchester University Press.

Martin, E. 1994. *Flexible Bodies: Tracking Immunity in American Culture from the Days of Polio to the Age of AIDS.* Boston, MA: Beacon Press.

McDowell, L. and Pringle, R. (eds). 1992. *Defining Women: Social Institutions and Gender Divisions.* Cambridge: Polity Press.

McHale, J.V. 2004. 'Regulating Genetic Databases: Some Legal and Ethical Issues' in S. Halliday and D.L. Steinberg (guest eds), *Medical Law Review* Special Issue 'The Regulated Gene: New Legal Dilemmas' 12(1): 70–96.

McKellen, I. 1993. 'Through a Gay Viewfinder', *Guardian*, 22 July, p. 20.

McNeil, M. 1987. *Gender and Expertise.* London: Free Association Books.

Meek, S. 1986. 'Reproductive Technology: Present Practices and Future Implications' in *Future Challenges for Australia: The Biotechology Revolution. Selected Papers.*

Merrick, J.C. and R.H. Blank (eds), 1993. *The Politics of Pregnancy: Policy Dilemmas in the Maternal-Fetal Relationship.* New York: Harrington Park Press.

Miller, D. et al. 1998. *The Circuit of Mass Communication: Media Strategies, Representation and Audience Reception in the AIDS Crisis.* Thousand Oaks, CA: Sage.

Minh-ha, T. 1989. *Woman, Native Other.* Bloomington: Indiana University Press.

Moore, S. 1993. 'Kiss and Don't Tell', *Guardian*, 23 July, p. 15.

Morgan. P. 1993. 'My Dinosaurs Scared the Life out of Di Says Spielberg', *Sun*, 17 July, p. 3.

Mort, F. 1987. *Dangerous Sexualities: Medico-moral Politics in England since 1830*. New York: Routledge and Kegan Paul.

Mulvey, L. 1988. 'Visual Pleasure and Narrative Cinema' in C. Penley (ed.), *Feminism and Film Theory*. London: Routledge.

Murray, S. and D. Holmes (eds). 2009. *Critical Interventions in the Ethics of Healthcare: Challenging the Principle of Autonomy in Bioethics*. Farnham: Ashgate.

Nair, J. 1992. 'Uncovering the *Zenana*: Visions of Indian Womanhood in Englishwomen's Writings 1813-1940', in C. Johnson-Odim (ed.), *Expanding the Boundaries of Women's History*. Bloomington: Indiana University Press.

Nead, L. 1988. *Myths of Sexuality*. Oxford: Blackwell.

Neale, S. 1990. 'Questions of Genre', *Screen* 31(1): 45–66.

Nelkin, D. and S. Lindee. 1995. *The DNA Mystique: The Gene as a Cultural Icon*. New York: W.H. Freeman (also published by the University of Michigan Press, 2004).

Nelkin, D. et al. 1994 [1991]. *Dangerous Diagnostics: Social Power of Biological Information*. Chicago: University of Chicago Press.

O'Riordan, K. 2010. *The Genome Incorporated: Constructing Biodigital Identity*. Farnham: Ashgate.

Oakley, A. 1976. 'Wisewoman and Medicine Man: Changes in the Management of Childbirth', in A. Oakley and J. Mitchell (eds), *The Rights and Wrongs of Women*. Harmondsworth: Penguin, pp. 17–45.

Parfitt, T. and Egorova, Y. 2005. *Genetics, Mass Media, and Identity: A Case Study of the Genetic Research on the Lemba and Bene Israel*. London: Routledge.

Parfitt, T. 2002. *The Lost Tribes of Israel: The History of a Myth*. London: Phoenix.

Parfitt, T. 2013. *Black Jews in Africa and the Americas*. New York: Harvard University Press.

Parmar, P. 1987. 'Hateful contraries: media images of Asian women', in R. Betterton (ed.), *Looking On: Images of Femininity in the Visual Arts and Media*. London: Pandora.

Parris, M. 1993. 'Genetic Genies Won't Go Back in the Bottle.' *The Times*, 17 July, p. 16.

Patton, C. 1985. *Sex and Germs: The Politics of AIDS*. Boston: South End Press.

Petchesky, R. Pollack. 1987. 'Fetal Images: The Power of Visual Culture in the Politics of Reproduction', *Feminist Studies* 13(2): 263–92.

Piercy, M. 1976. *Woman on the Edge of Time*. New York: Alfred A. Knopf.

Piercy, M. 1993. *Body of Glass*. Harmondsworth: Penguin.

Piller, C. and K.R. Yamamoto. 1988. *Gene Wars: Military Control over the New Genetic Technologies*. New York: Beech Tree Books, William Morrow.

Plummer, K. 1995. *Telling Sexual Stories: Power, Change and Social Worlds*. London: Routledge.

Proctor, R.N. 1988. *Racial Hygiene: Medicine and the Nazis*. Cambridge, MA: Harvard University Press.

Radford, T. 1993. 'Your Mother Should Know',*Guardian*, 17 July, p. 1.

Radford, T. 1993. 'Code of Conduct', *Guardian*, 21 July, p. 2.

Rajan, K.S. 2007. *Biocapital: The Constitution of Postgenomic Life*. Durham, NC: Duke University Press.

Redman, P. 1997. 'Invasion of the Monstrous Others: Heterosexual Masculinities, the "AIDS Carrier" and the Horror Genre' in D.L. Steinberg et al. (eds), *Border Patrols: Policing the Boundaries of Heterosexuality*. London: Cassell, pp. 98–117.

Redman, P. 1998. 'Narrative and the Production of Subjectivity', in *Boys in Love: Narrative, Genre and the Production of Heterosexual Masculinities*. Unpublished PhD Thesis, School of English, University of Birmingham. England.

Reissman, C.K. 1992. 'Women and Medicalisation: A New Perspective', in G. Kirkup and L. Smith Keller (eds), *Inventing Women: Science, Technology and Gender*. Cambridge: Polity Press.

Ridley, M. 2000. *Mendel's Demon: Gene Justice and the Complexity of Life*. London: Weidenfeld & Nicolson.

Roberts, D. 1992. *Killing the Black Body: Race, Reproduction, and the Meaning of Liberty*. New York: Vintage.

Rose, N. 2007. *The Politics of Life Itself: Biomedicine, Power and Subjectivity in the Twenty-First Century*. Princeton, NJ: Princeton University Press.

Rose, S. and H. Rose. 2012. *Genes, Cells and Brains: Bioscience's Promethean Promises*. London: Verso.

Rose, S. et al. 1984. *Not in Our Genes*. New York: Pantheon Books.

Rosoff, B. and E. Tobach (eds). 1994. *Challenging Racism and Sexism: Alternatives to Genetic Explanations – Genes & Gender VII*. New York: The Feminist Press.

Roth, R. 1993. 'At Women's Expense: The Costs of Fetal Rights' in Janna C. Merrick and Robert H. Blank. (eds), *The Politics of Pregnancy: Policy Dilemmas in the Maternal-Fetal Relationship*. Binghampton, NY: Harrington Park Press, pp. 117–36.

Rothman, B. Katz. 1987. *The Tentative Pregnancy: Prenatal Diagnosis and the Future of Motherhood*. New York: Penguin.

Rothman, B. Katz. 1998. *Genetic Maps and Human Imaginations: The Limits of Science in Understanding Who We Are*. New York: W.W. Norton & Company.

Rutter, Sir M. 1996. 'Introduction: Concepts of Antisocial Behaviour, of Cause and of Genetic Influences' in CIBA Foundation Symposium 194, *Genetics of Criminal and Antisocial Behaviour (Symposium 194)*. Chichester: John Wiley & Sons.

Sargisson, L. 1996. *Contemporary Feminist Utopianism*. London: Routledge.

Scannell, P. and D. Cardiff. 1981. 'Serving the Nation: Public Service Broadcasting Before the War' in B. Wites, T. Bennett and G. Martin (eds), *Popular Culture: Past and Present*. Milton Keynes: Open University Press, pp. 161–87.

Scannell, P. 1990. 'Public Service Broadcasting: the History of a Concept' in A. Goodwin and G. Whannel (eds), *Understanding Television*. London: Routledge, pp. 11–29.

Schiebinger, L. 1993. *Nature's Body: Gender in the Making of Modern Science*. Boston, MA: Beacon Press.

Schiebinger, L. 2004. 'Feminist History of Colonial Science', *Hypatia* 19(1): 233–54.

Schneps, L. and C. Comez. 2013. *Math on Trial: How Numbers get Used and Abused in the Courtroom*. New York: Basic Books.

Science and Technology Sub-Group, 1991. 'In the Wake of the Alton Bill: Science, Technology and Reproductive Politics' in S. Franklin et al. (eds), *Off Centre – Feminism and Cultural Studies*. London: HarperCollins.

Sedgwick, E. Kosofky. 1993. *Tendencies*. Durham, North Carolina: Duke University Press.

Shatz, D. 2004. *Peer Review: A Critical Inquiry*. Lanham, MD: Rowman and Littlefield.

Shildrick, M. and J. Price. 2008. *Vital Signs: Feminist Reconfigurations of the Biological Body*. Edinburgh: University of Edinburgh Press.

Shildrick, M. 1997. *Leaky Bodies and Boundaries: Feminism, Postmodernism and (Bio)ethics*. London: Routledge.

Showalter, E. 1992. *Sexual Anarchy: Gender and Culture at the* Fin de Siècle. London: Virago.

Silberg, J. et al. 1996. 'Heterogeneity Among Juvenile Antisocial Behaviours: Findings from the Virginia Twin Study of Adolescent Behavioural Development' in CIBA Foundation Symposium 194, *Genetics of Criminal and Antisocial Behaviour (Symposium 194)*. Chichester: John Wiley & Sons.

Smyth, L. 1999. 'Abortion and Family Values: The X Case, Sexuality and "Irishness"' in D. Epstein and J.T.A. Sears (eds), *Dangerous Knowing: Sexuality, Pedagogy and Popular Culture*. London: Cassell.

Smyth, L. 2005. *Abortion and Nation: The Politics of Reproduction in Contemporary Ireland*. Aldershot: Ashgate.

Solomos, J. 1989. *Race and Racism in Contemporary Britain*. London: Macmillan.

Sonczewski, J. 1986. *A Door into Ocean*. New York: Avon.

Sontag, S. 1977. [1991 edition]. *Illness as Metaphor/AIDS and its Metaphors*. London: Penguin.

Spallone, P. 1992. *Generation Games: Genetic Engineering and the Future for our Lives*. London: The Women's Press.

Spallone, P. 1989. *Beyond Conception: The New Politics of Reproduction*. Women in Society. London: Macmillan.

Spanier, B.B. 1995. *Im/Partial Science: Gender Ideology in Molecular Biology*. Bloomington: Indiana University Press.

Stacey, J. 1991. 'Promoting Normality: Section 28 and the Regulation of Sexuality' in S. Franklin, C. Lurie and J. Stacey (eds), *Off-Centre: Feminism and Cultural Studies*. London: Routledge, pp. 284–304.

Stacey, J. 2010. *The Cinematic Life of the Gene*. Durham, NC: Duke University Press.

Steinberg, D.L.. 1996. 'Cultural Regimes of the Body: Introduction' in D.L. Steinberg (guest editor), 'Cultural Regimes of the Body' (Special Issue) *Women: A Cultural Review* 7(3): 225–8.

Steinberg, D.L. 1991. 'Adversarial Politics: The Legal Construction of Abortion' in S. Franklin, C. Lurie and J. Stacey (eds), *Off-Centre: Feminism and Cultural Studies*. London: Routledge, pp. 175–89.

Steinberg, D.L. 1997. 'Technologies of Heterosexuality: Eugenic Reproductions Under Glass' in D.L. Steinberg, D. Epstein and R. Johnson (eds)*, Border Patrols: Policing the Boundaries of Heterosexuality*. London: Cassell.

Steinberg, D.L. 1997. *Bodies in Glass: Genetics, Eugenics, Embryo Ethics*. Manchester: Manchester University Press.

Steinberg, D.L. 2000. 'Bodies of Knowledge: Shifting Iconographies of Maternal-Fetal Conflict', *Feminist Review* (Spring): 139–43.

Steinberg, D.L., D. Epstein and R. Johnson (eds). 1997. *Border Patrols: Policing the Boundaries of Heterosexuality*. London: Cassell.

Steinem, G. 1984. *Outrageous Acts and Everyday Rebellions*. London: Fontana.

Stepan, N. Leys. 1993. 'Race and Gender: The Role of Analogy in Science' in S. Harding (ed.), *The 'Racial' Economy of Science: Toward a Democratic Future*. Bloomington. Indiana University Press, pp. 359–76.

Stoppard, T. 1986. *Jumpers*. London: Faber & Faber.

Stott, R. 1989. 'The Dark Continent; Africa as Female Body in Haggard's Adventure Fiction', *Feminist Review* 32: 69–89.

Strathern, M. 1992. *Reproducing the Future: Anthropology, Kinship and the New Reproductive Technologies*. Manchester: Manchester University Press.

Strobel, M. 1991. *European Women and the Second British Empire*. Bloomington: Indiana University Press.

Sunder, R.K. 2006. *Biocapital: The Constitution of Postgenomic Life*. Durham, NC and London: Duke University Press.

Swain, G. 1993. 'Men Inherit Gay Genes from Mum', *Daily Mirror*, 17 July, p. 9.

Taylor, B. 1993. "Unconsciousness and Society: The Sociology of Sleep," *International Journal of Politics, Culture and Society* 6(3): 463–71.

Taylor, P. et al. (eds). 1997. *Changing Life, Genomes, Ecologies, Bodies, Commodities*. Minneapolis: University of Minnesota Press.

Terry, J. and J. Urla (eds). 1995. *Deviant Bodies*. Bloomington: Indiana University Press.

Terry, J. 1997. 'The Seductive Power of Science in the Making of Deviant Subjectivity' in V.A. Rosario (ed.), *Science and Homosexualities*. New York: Routledge.

Thacker, E. 2005. *The Global Genome: Biotechnology, Politics and Culture*. Cambridge: MA: MIT Press.

Thompson, E.P. 1991. *The Making of the English Working Class*. London: Penguin.

Throsby, K. 2004. *When IVF Fails: Feminism, Infertility and the Negotiation of Normality*. Basingstoke: Palgrave.

Throsby, K. 2006. 'The unaltered body? Rethinking the body when IVF fails', *Science Studies* 19(2): 77–97.

Throsby, K. and R. Gill. 2004. '"It's different for men": Masculinity and IVF'. *Men and Masculinities* 6(4): 330–48.

Throsby, K. and C. Roberts. 2010. 'Getting bigger: children's bodies, genes and environments' in S. Parry and J. Dupré (eds), *Nature after the Genome*. Malden, MA: Wiley-Blackwell.

Throsby, K. 2010. '"Doing what comes naturally…" Negotiating normality in accounts of IVF-failure' in L. Reed and P. Saukko (eds), *Governing the Female Body: Gender, Health, and Networks of Power*. Albany, NY: State University of New York Press, pp. 233–52.

Times, The. 1993. 'Homosexuals and a Tolerant Society' (Letters to the Editor), 20 July, p. 17.

Times., The. 1993. 'Genes and the Man: Scientific Discoveries do not Resolve Moral Dilemnas', 17 July, p. 17.

Tobach, E. and B. Rosoff. 1994. *Challenging Racism and Sexism: Alternatives to Genetic Determinism*. Bloomington, IN: Indiana University Press.

Tudge, C. 2000. *In Mendel's Footnotes: An Introduction to the Science and Technologies of Genes and Genetics from the 196h Century to the 22nd*. London: Jonathan Cape.

Turney, J. 1998. *Frankenstein's Footsteps: Science, Genetics and Popular Culture*. New Haven, CT: Yale University Press.

Usher, J. 1991. *Women's Madness: Misogyny or Mental Illness?* London: Harvester Wheatsheaf.

Van Dijck, J. 1998. *Imagenation: Popular Images of Genetics*. Basingstoke: Macmillan.

Vermeulen, N, S. Tamminen, and A. Webster (eds). 2012. *Bio-Objects: Life in the 21st Century*. Farnham: Ashgate.

Waldby, C. 1996. *AIDS and the Body Politic: Biomedicine and Sexual Difference*. London: Routledge.

Walkowitz, J.R. 1980. *Prostitution and Victorian Society: Women, Class and the State*. Cambridge: Cambridge University Press.

Watney, S. 1988. 'AIDS, "Moral Panic" Theory and Homophobia' in Peter Aggleton and Hilary Homans (eds), *Social Aspects of AIDS*. Philadelphia: Falmer Press.

Watson, P. 1993. 'Mums Pass on Gay Gene to Sons say Doctors'. *Sun*, 17 July, p. 6.

Weatherall, D.J. et al. 1986. 'Analysis of Foetal DNA for the Diagnosis and Management of Genetic Disease' in The CIBA Foundation (ed.), *Human Embryo Research: Yes or No?* London: Tavistock, pp. 83–99.

Weiss, S.F. 1987. *Race Hygiene and National Efficiency: The Eugenics of William Schallmayer*. Berkeley, CA: University of California Press.

Whitelegg, L. 1992. 'Girls in Science Education: of Rice and Fruit Trees' in G. Kirkup and L. Smith Keller (eds), *Inventing Women: Science, Technology and Gender*. London: Polity Press, pp. 178–87.

Whitfield, M. 1993. 'My Fear is Having Straight Children'. *Independent*, 17 July, p. 1.

Wieseltier, L. 2013. 'Leon Wieseltier Responds to Steven Pinker's Scientism'. newrepublic.com <http://www.newrepublic.com/article/114548/leon-wieseltier-responds-steven-pinkers-scientism>.

Williams, S. 2005. *Sleep and Society: Sociological Ventures Into the (Un)known*. New York: Taylor and Francis.

Williamson, J. 1997. 'Saving Grace' in *Guardian* (Weekend Magazine), 30 August, p. 6.

Williamson, R. 1986. 'Research Needs and the Reduction of Severe Congenital Disease' in The CIBA Foundation (ed.), *Human Embryo Research: Yes or No?* pp. 105–14. London: Tavistock.

Winston, R. 1997. *The Future of Genetic Manipulation*. London: Orion.

Winston, R. 1999. *The IVF Revolution*. London: Vermilion Publishing.

Wites, B., T. Bennett and G. Martin. 1981. *Popular Culture: Past and Present*. Milton Keynes: Open University Press.

Witz, A. 1992. *Professions and Patriarchy*. London: Routledge.

Woolfson, A. 2000. *Life Without Genes*. London: HarperCollins.

Yanchinski, S. 1985. *Setting Genes to Work: The Industrial Era of Biotechnology*. Harmondsworth: Penguin.

Zakin, E. 1999. 'Catherine Vassellou: Textures of Light; Vision and Touch in Irigaray, Levinas and Merleau-Ponty'. *APA NewsLetter on Feminism* 99(1): 58–60.

Online Media Sources

Adams, Tim. 12 May 2013. 'How to spot a murderer's brain'. *Observer* <http://www.guardian.co.uk/science/2013/may/12/how-to-spot-a-murderers-brain>.

Bergman, Lowell and Andres Cedial (prod.) 2012. 'The Real CSI *Frontline* (PBS) Transcript' <http://www.pbs.org/wgbh/pages/frontline/criminal-justice/real-csi/transcript-18/>.

Colata, Gina. 25 December 2012. 'Seeking Answers in Genome of Gunman'. *The New York Times* <http://www.nytimes.com/2012/12/25/science/scientists-to-seek-clues-to-violence-in-genome-of-gunman-in-newtown-conn.html?hpw>.

Colata, Gina. 6 September 2012 'Far From "Junk," DNA Dark Matter Proves Crucial to Health'. *The New York Times* <http://www.nytimes.com/2012/09/06/science/far-from-junk-dna-dark-matter-proves-crucial-to-health.html?_r=1>.

Collins, Ellen. 14 August 2013. 'Open Access Isn't Enough in Itself'. *Guardian* <http://www.theguardian.com/higher-education-network/blog/2013/aug/14/open-access-media-coverage-research>.

Connor, Steven. 1 November 1995. 'The 'gay gene' is back on the scene. Does new research finally prove that homosexuality can be inherited? Steve Connor investigates the latest controversial findings'. *Independent* <http://www.independent.co.uk/news/the-gay-gene-is-back-on-the-scene-1536770.html>.

Dodd, Vikram. 20 May 2013. 'Police retain DNA from thousands of children'. *Guardian* <http://www.guardian.co.uk/uk/2013/may/20/police-retain-dna-thousands-children>.

Donaldson James, Susan. 16 August 2013. '8 Year Old Never Ages, Could Reveal "Biological Immortality"' <http://abcnews.go.com/Health/girl-ages-unravel-secret-eternal-youth/story?id=19974247>.

Dr Psyphago. 2 November 2012. 'Gene for poor science journalism discovered'. Collectivelyunconscious.wordpress.com <https://collectivelyunconscious.wordpress.com/2012/11/02/gene-for-poor-science-journalism-discovered/>.

Drumming, Neil. 26 September 2013. 'The Good Wife Gears Up for Battle' Salon <http://www.salon.com/2013/09/25/the_good_wife_gears_up_for_battle/>.

Ellenberg, Jordan. 12 June 2013. 'Doubt and the Double Helix'. *Slate* <http://www.slate.com/articles/technology/technology/2013/06/dna_math_if_police_find_a_genetic_match_that_doesn_t_mean_they_have_the.html>.

Entine, John. 8 May 2012. 'Jews Are a "Race, Genes Reveal'. forward.com <http://forward.com/articles/155742/jews-are-a-race-genes-reveal/?p=all>.

Fanelli, Daniele and John P.A. Ioannidis. 2013. 'US studies may overestimate effect sizes in softer research'. Proceedings of the National Academy of Sciences of the United States of America. pnas.org <http://www.pnas.org/content/110/37/15031>.

Flatow, Nicole. 31 July 2013. 'Court Clerk Fired For Helping Secure DNA Test That Proved A Man's Innocence'. thinkprogress.org <http://thinkprogress.org/justice/2013/07/30/2384751/court-clerk-fired-for-helping-secure-dna-test-that-proved-a-mans-innocence/>.

Gautam, Naik and Robert Lee Holtz. 6 September 2012. 'Junk DNA Debunked: Studies Find Human Genomic Makeup is Vastly Messier; New Disease Links Seen.' wsj.com <http://online.wsj.com/article/SB10000872396390443589304577633560336453228.html>.

Goldstein, Joseph. 27 January 2013. '3 Years After Inception, a DNA Technique Yields Little Success for the Police'. *The New York Times* <http://www.nytimes.com/2013/01/28/nyregion/partial-match-dna-technique-has-

yielded-little-success-for-police.html?_r=0&adxnnl=1&emc=eta1&adxnn
lx=1381998310-8mst6TnJJeiOemG5KEXaWw>.

Good Morning America. 16 August 2013. '8 Year Old Never Ages, Could Reveal "Biological Immortality"'. *ABC* <http://abcnews.go.com/Health/girl-ages-unravel-secret-eternal-youth/story?id=19974247>.

Gray, Richard. 12 September 2013. 'Jurassic Park Ruled Out: dinosaur DNA could not survive in Amber'. *The Telegraph* <http://www.telegraph.co.uk/science/dinosaurs/10303795/Jurassic-Park-ruled-out-dinosaur-DNA-could-not-survive-in-amber.html>.

Guardian. 31 July 2012. 'Barack Obama may be descendant of first African slave in colonial America'. *Guardian* <http://www.guardian.co.uk/world/us-news-blog/2012/jul/30/barack-obama-descendant-first-african>.

Guardian. 11 September 2012. 'Woolly Mammoth Remains may Contain Living Cells'. *Guardian* <http://www.guardian.co.uk/science/2012/sep/11/wooly-mammoth-remains-living-cells>.

Guardian. 12 July 2013. 'DNA testing of Albert DeSalvo may lay Boston Strangler identity to rest'. *Guardian* <http://www.guardian.co.uk/world/2013/jul/12/boston-strangler-desalvo-identity-dna>.

Innocence Project, The <http://www.innocenceproject.org>.

Jha, Alok. 26 August 2013. 'Research finds "US effect" exaggerates results in human behaviour studies'. *Guardian* <http://www.theguardian.com/science/2013/aug/26/research-funding-exaggerates-results>.

Jha, Alok. 26 August 2013. 'Research finds "US effect" exaggerates results in human behaviour studies Confirmation bias'. *Guardian* <http://www.theguardian.com/science/2013/aug/26/research-funding-exaggerates-results>.

Jha, Alok. 5 September 2012 'Breakthrough study overturns theory of "junk DNA" in genome'. *Guardian* <http://www.guardian.co.uk/science/2012/sep/05/genes-genome-junk-dna-encode>.

Johnstone, Ian. 8 August 2013. 'FBI re-opens 1964 stolen baby case after DNA proves wrong child was returned'. usnews.nbcnews.com <http://usnews.nbcnews.com/_news/2013/08/08/19927561-fbi-re-opens-1964-stolen-baby-case-after-dna-proves-wrong-child-was-returned>.

Jolie, Angelina. 14 May 2013. 'My Medical Choice'. *The New York Times* <http://www.nytimes.com/2013/05/14/opinion/my-medical-choice.html?_r=0>.

Jones, Thomas. 10 April 2010. 'The rise of DNA analysis in crime solving'. *Guardian* <http://www.theguardian.com/politics/2010/apr/10/dna-analysis-crime-solving>.

Kean, Sam. 3 October 2013. 'Is Junk DNA Really Junky?'. *Slate* <http://www.slate.com/articles/health_and_science/human_genome/2013/10/junk_dna_debate_critics_of_the_encode_project_cite_humped_bladderwort.html?wpisrc=burger_bar>.

Kedmey, Dan. 27 August 2013. 'What DNA Testing Reveals About India's Caste System'. world.time.com <http://world.time.com/2013/08/27/what-dna-testing-reveals-about-indias-caste-system/>.

Konnikova, Maria. 15 May 2013. 'Angelina Jolie, Meet Nate Silver'. *Slate* <http://www.salon.com/2013/05/15/angelina_jolie_meet_nate_silver/>.

Lai, Jennifer. 28 June 2013. 'Britain Is Moving Closer to Having "Three-Parent" IVF Babies'. *Slate* <http://www.slate.com/blogs/the_slatest/2013/06/28/united_kingdom_may_approve_creating_babies_with_dna_from_3_people_thanks.html>.

McKie, Robin. 24 February 2013. 'Scientists attacked over claim that "junk DNA" is vital to life'. *Guardian* <http://www.guardian.co.uk/science/2013/feb/24/scientists-attacked-over-junk-dna-claim>.

National Human Genome Research Institute (USA). 'Learning About Tay Sachs Disease'. www.genome.gov <http://www.genome.gov/10001220>.

Nugent, Helen. 10 March 2012. 'Paedophile who abducted girls 30 years ago caught by DNA'. *Guardian* <http://www.guardian.co.uk/uk/2012/mar/09/paedophile-dna-david-bryant>.

Obasogie, Osagie K. 25 July 2013. 'High Tech, High Risk Forensics'. *The New York Times* <http://nyti.ms/14409rD>.

Palmer, Brian. 11 March 2007. 'Is Criminality Genetic?' *Slate* <http://www.slate.com/articles/health_and_science/explainer/2013/03/osama_bin_laden_son_in_law_does_criminality_run_in_families.html?wpisrc=sl_ipad>.

Palmer, Brian. 30 September 2013. 'Where Are All the Miracle Drugs?' *Slate* <http://www.slate.com/articles/health_and_science/human_genome/2013/09/human_genome_drugs_where_are_the_miracle_cures_from_genomics_did_the_genome.html>.

Palmer, Brian. 30 September 2013. 'Where Are All the Miracle Drugs?' *Slate* <http://www.slate.com/articles/health_and_science/human_genome/2013/09/human_genome_drugs_where_are_the_miracle_cures_from_genomics_did_the_genome.html>.

Pauwels, Eleonor. 12 June 2013. 'Watch Where You Shed Your DNA—an Artist Might Use It'. *Slate* <http://www.slate.com/blogs/future_tense/2013/05/31/heather_dewey_hagborg_artist_creates_3_d_models_of_people_using_found_genetic.html>.

Peters, Justin. 12 July 2013. 'In 2001, DNA Evidence Didn't Point to Albert DeSalvo as the Boston Strangler. New DNA Tests Do. What Gives?' *Slate* <http://www.slate.com/blogs/crime/2013/07/11/albert_desalvo_mary_sullivan_new_evidence_in_boston_strangler_case_contradicts.html>.

reason.com. 15 May 2013. 'Angelina Jolie's Genetic Self-Ownership Is the Future of Medicine'. <http://reason.com/blog/2013/05/14/angelina-jolies-genetic-self-ownership-i>.

Rettner, Rachael. 23 February 2012. 'New Breast Cancer Gene Found'. livescience. com <http://www.livescience.com/18607-breast-cancer-gene.html>.

Rivlin-Nadler, Max. 18 August 2013. 'Could this 8 Year Old be the Secret to Immortality?' *Gawker* <http://gawker.com/could-this-8-year-old-be-the-key-to-immortality-1163129595>.

Sample, Ian. 31 July 2013. 'Woolly Mammouth DNA may lead to a Resurrection of the Ancient Beast'. *Guardian* <http://www.guardian.co.uk/science/2012/sep/11/wooly-mammoth-remains-living-cells>.

Sample, Ian. 7 July 2013. 'IVF baby born using revolutionary genetic-screening process'. *Guardian* <http://www.guardian.co.uk/science/2013/jul/07/ivf-baby-born-genetic-screening>.

Sample, Ian. 28 June 2013. 'Three Person IVF: UK Government Backs Mitochondrial Transfer'. *Guardian* <http://www.guardian.co.uk/science/2013/jun/28/uk-government-ivf-dna-three-people>.

Sample, Ian. 17 May 2013. 'IVF Could be Revolutionised by New Technique, Clinic Says'. *Guardian* <http://www.theguardian.com/society/2013/may/17/ivf-revolutionised-new-technique-clinic>.

Stern, Michael Joseph. 28 June 2013. 'Born This Way? Scientists may have Found a Biological Basis for Homosexuality. That Could be Bad News for Gay Rights'. *Slate* <http://www.slate.com/articles/health_and_science/science/2013/06/biological_basis_for_homosexuality_the_fraternal_birth_order_explanation.html>.

West, Lindy. 28 December 2012. 'Geneticists Plan to Examine Adam Lanza's DNA for "Abnormalities"'. Jezebel <http://jezebel.com/5971611/geneticists-plan-to-examine-adam-lanzas-dna-for-abnormalities>.

Wilcox, Christine. 11 September 2012. 'Scientists play a large role in bad medical reporting'. scientificamerican.com <http://blogs.scientificamerican.com/science-sushi/2012/09/11/scientists-play-a-large-role-in-bad-medical-reporting/>.

Zimmer, Carl. 17 September 2013. 'DNA Double Take' nytimes.com <http://nyti.ms/187ivox>.

Zuger, Abigail. 29 February 2012 'Genomics as a Final Frontier, or Just a Way Station.' 3quarksdaily.com <http://www.3quarksdaily.com/3quarksdaily/2012/02/genomics-as-a-final-frontier-or-just-a-way-station.html>.

Radio, Film and Television

BBC Radio 4. 1991. 'The Language of the Genes' *Reith Lectures* (presented by Dr Steve Jones) <http://www.bbc.co.uk/programmes/p00gq073>.

BBC Radio 4. 20 August 2013. *Raising Allosaurus: The Dream of Jurassic Park* <http://www.bbc.co.uk/programmes/b038c7b9>http://www.bbc.co.uk/news/science-environment-23602142>.

Bergman, L. and A. Cedial (prod.). 2012. 'The Real CSI: *Frontline*'. PBS <http://video.pbs.org/video/2223977258/>.

CSI. 2005. 'Unbearable', Season 5, Episode 14.

CSI. 2002. 'The Execution of Catherine Willows', Season 2, Episode 6.

CSI. 2011. 'Targets of Obsession', Season 11, Episode 15.

CSI: Miami. 2010. 'Out of Time', Season 8, Episode 1.

Good Morning America. August 2009. 'The Amazing Girl Who Doesn't Age'. *ABC* <https://www.youtube.com/watch?v=6gF9SkmCEEU>.

Hale, C. (dir.). 1999. *To the Ends of the Earth: Search for the Sons of Abraham.* Channel 4 / NOVA (UK).

Holt, S. and N. Patterson (dirs). 2007. *Ghost in Your Genes.* NOVA (screened on PBS, USA) <http://www.pbs.org/wgbh/nova/genes/>.

Minh-ha, Trinh (dir.). 1982. *Reassemblage.*

NOVA. 22 February 2000. 'Tudor Parfitt's Remarkable Quest' <http://www.pbs.org/wgbh/nova/israel/parfitt.html>.

Onion, The. 25 September 2013. 'DNA Evidence Frees Black Man Convicted Of Bear Attack'. YouTube <http://m.youtube.com/watch?feature=relmfu&v=u1cgHEWG-BA>.

Slate.V. Staff. 14 August 2013. 'DNA May Finally Solve Mona Lisa Mystery' (VIDEO). *Slate* <http://www.slate.com/blogs/trending/2013/08/14/mona_lisa_identity_dna_evidence_may_confirm_lisa_gherardini_inspired_the.html>.

Spielberg, S. (dir.). 1981. *Raiders of the Lost Ark.*

Spielberg, S. (dir.). 1993. *Jurassic Park.*

Index